高等院校新能源专业系列教材
普通高等教育新能源类"十四五"精品系列教材

融合教材

Solar Energy Professional English

太阳能专业英语

古丽米娜 等 主编

中国水利水电出版社
www.waterpub.com.cn
·北京·

内 容 提 要

本书主要介绍太阳能光电转换基本原理、各类型太阳电池材料与器件基础知识及其光电性能、光伏组件及光伏系统的应用，太阳能光热利用基本原理、不同类型光热发电系统及集热器、太阳能加热及冷却系统、太阳能储热材料等内容。本书结合太阳能发电的基本原理、技术利用方式以及工程应用的进展，引出所需掌握的太阳能专业英语核心知识，为进一步提高读者的太阳能专业英语阅读、写作能力奠定坚实的基础。

本书既可作为太阳能专业人员的参考资料，也可作为相关专业的教学用书。

图书在版编目（CIP）数据

太阳能专业英语 / 古丽米娜等主编. -- 北京 : 中国水利水电出版社, 2022.11
高等院校新能源专业系列教材　普通高等教育新能源类"十四五"精品系列教材
ISBN 978-7-5226-1053-5

Ⅰ. ①太… Ⅱ. ①古… Ⅲ. ①太阳能—英语—高等学校—教材 Ⅳ. ①TK511

中国版本图书馆CIP数据核字(2022)第196855号

书　　名	高等院校新能源专业系列教材 普通高等教育新能源类"十四五"精品系列教材 **太阳能专业英语** TAIYANGNENG ZHUANYE YINGYU
作　　者	古丽米娜　等　主编
出版发行	中国水利水电出版社 （北京市海淀区玉渊潭南路1号D座　100038） 网址：www.waterpub.com.cn E-mail: sales@mwr.gov.cn 电话：(010) 68545888（营销中心）
经　　售	北京科水图书销售有限公司 电话：(010) 68545874、63202643 全国各地新华书店和相关出版物销售网点
排　　版	中国水利水电出版社微机排版中心
印　　刷	天津嘉恒印务有限公司
规　　格	184mm×260mm　16开本　10印张　316千字
版　　次	2022年11月第1版　2022年11月第1次印刷
印　　数	0001—3000册
定　　价	52.00元

凡购买我社图书，如有缺页、倒页、脱页的，本社营销中心负责调换

版权所有·侵权必究

前 言

随着环境状况的日益恶化、化石能源的逐渐枯竭，新能源、新材料的开发及应用已成为全球的首要课题。其中，太阳能发电材料、技术应用及产业发展是重点之一。因此，在全球大力发展新能源尤其是光伏产业之际，掌握太阳能发电基础知识及相关信息非常重要，将其作为高等基础教育的一部分也尤为关键。与此同时，英语作为一种重要的全球化交流工具，在国际交流合作中也发挥着非常重要的作用。因此，学好太阳能专业英语，是学生、学者和工程技术人员与国际太阳能同行交流，积极获取国际太阳能科研信息，掌握国外太阳能学科发展动态，参加国际学术交流的前提。为此，本书提供了一些需要掌握的太阳能英语核心知识，希望能对各阶段相关人员专业英语水平的提高有所帮助。

本书整体上由浅入深地介绍了太阳能发电材料、发电技术应用的方方面面。每一章均由该章节相关的专业知识内容、专业词汇、练习题、课后延伸阅读等部分组成，且每章的课后阅读紧密围绕该章节所涉及的知识，并在其基础上进一步扩展相关学术研究内容，使得阅读者能更广泛地接触到相关的学术前沿知识。本书共分为两部分，11章，其中第一部分主要介绍光伏发电相关的基础知识、各类光伏发电材料及光伏电池、光伏发电技术的应用；第二部分主要介绍太阳能光热技术的基本原理、各类光热发电系统及集热器、光热加热冷却系统、太阳能储热材料及光热技术在各领域的应用等。整书内容涵盖了从基础知识到应用技术，以及新型太阳能发电材料的内容。

本书在帮助学生掌握太阳能发电材料、技术应用相关知识的基础上，进一步提高学生的专业英语阅读能力，拓展和深化学生对太阳能发电技术的认知。本书是为培养学生应用能力而编写的基础专业英语教材，其特点如下：

（1）针对性强。本书内容贴合太阳能发电材料、技术应用相关专业知识，并力求知识面更周全、更准确。

（2）内容全面。本书各章节内容从太阳能发电基本原理、发电材料种类到太阳能发电技术应用方式等，从太阳能学科最基础的知识内容延伸到最终的应用体系，涉及内容全面，涵盖面广，并通过课后阅读，进一步扩展读者的知识面。

（3）词汇丰富。本书所选词汇几乎涵盖了从太阳能基础知识到应用的绝大多数专业英语词汇，并且单独列出了中英文对照的部分，因此既可以作为高等学校教材，又可以作为各行业读者的自学教材。

（4）具备知识扩展内容。本书通过练习题以及课后延伸阅读等内容的添加，为读者进一步对所学章节进行思考、延伸扩展所学内容奠定了良好的基础，为高等学校学生、工程技术人员在学习之后思维的延伸、了解科研前沿动态提供了条件。

本书共 11 章，由古丽米娜统筹规划，并担任主编，张灿灿担任副主编。由于本书作者水平有限，书中难免有不足和疏漏之处，恳请各位专家、同仁和广大读者批评指正。

编者
2022 年 10 月

Contents

前言

Solar Energy 1

Part Ⅰ Photovoltaic

Chapter 1

Photovoltaic overview 7

1.1　Photovoltaic (PV) history 7
1.2　Status of photovoltaics 8
1.3　Applications of photovoltaics 12
1.4　Technology trends 21

Chapter 2

Principle of photovoltaic 23

2.1　P-n Junction of the solar cells 23
2.2　Solar cell types 25
2.3　The structure and working mechanism of the solar cells ... 28
2.4　The fundamental characteristics of the solar cells 31

Chapter 3

Solar cells 33

3.1　Silicon-based solar cells 33
3.2　Inorganic compound solar cells 43
3.3　Novel organic semiconductor thin film solar cells 49
3.4　Concentrator solar cells 60

Chapter 4
PV Modules ... 64
4.1 The definition and types of PV modules ... 64
4.2 Module structure ... 65
4.3 Encapsulation technology ... 67

Chapter 5
PV Systems ... 69
5.1 PV systems and types ... 69
5.2 Stand-alone PV systems ... 70
5.3 Grid connected PV systems ... 72
5.4 Hybrid power systems ... 73

Part II Solar Thermal Energy

Chapter 6
Solar thermal energy overview ... 79
6.1 The brief history of solar thermal energy ... 79
6.2 Solar thermal power generation system ... 80

Chapter 7
Solar thermal power generation system ... 83
7.1 Brief introduction of solar thermal power generation ... 83
7.2 Parabolic trough solar thermal power system ... 84
7.3 Tower solar thermal power system ... 85
7.4 Dish solar thermal power system ... 87
7.5 Linear Fresnel solar thermal power system ... 88

Chapter 8
Solar heating and cooling system ... 91
8.1 Solar heating system ... 91
8.2 Passive solar heating system ... 93
8.3 Solar cooling system ... 98

Chapter 9
Non-concentrating solar collector ... 102
9.1　Flat plate collector ... 102
9.2　Evacuated tube solar collector ... 110

Chapter 10
Solar thermal energy storage materials ... 115
10.1　Overview of solar thermal energy storage materials ... 115
10.2　Sensible heat storage material ... 116
10.3　Phase change materials ... 118
10.4　Heat storage by chemical reaction ... 122

Chapter 11
Practical application of solar thermal technology ... 124
11.1　Application of low temperature solar energy ... 125
11.2　Application of medium temperature solar energy ... 126

Appendix ... 130

References ... 147

Solar Energy

Solar energy overview

Solar energy is an inexhaustible clean energy, completely different from the traditional energy sources such as coal and oil. The development and utilization of solar energy have been integrated into people's production and life, and have been valued by countries around the world. The study of the basic knowledge, resource distribution, conversion and utilization of solar energy is of great significance to the efficient use of solar energy.

The use of solar energy will not lead to "greenhouse effect", nor will it pollute the environment. Solar energy is available locally, convenient and safe, requiring no transportation. Nevertheless, due to the dispersive, intermittent, and regional availability, the amount of solar radiation received by different regions varies considerably. Therefore, when studying the distribution of the solar energy resources, people mainly consider their abundance, which in turn can be examined by the total amount of solar radiation and the total solar sunshine hours.

Under normal circumstances, from the desert zone with more sunshine to the polar regions, the global annual cumulative sunshine can generally reach $640 \sim 2300$ kWh/m^2 or more. The region with the optimum solar energy resources in China is the Qinghai – Tibet Plateau, which is comparable to the best areas in India and Pakistan. Solar energy is abundant, but the use of it is less than 1/1000 of the exploitable amount; therefore, its development in China is worth expecting. For solar irradiation, most of the country is in the medium – high radiation area, except Guizhou Plateau and Chongqing. In general, the amount of solar radiation is uncertain, resulting in its unstable availability.

China is in the northern hemisphere, and generally in the mid – latitude area, and the solar energy resource is rich with the annual average solar irradiation of about 5.9 kJ/m^2. The solar radiation of the valley areas in Tibet is as high as $7.5 \sim 7.9$ kJ/m^2, areas with the lowest sunshine such as Chongqing and Sichuan Basin can also receive the solar radiation of $3.3 \sim 4.2$ kJ/m^2 (Table 1).

Table 1　　　　Annual radiation of different solar energy resource belts in China

Resource belt	Classification	Annual radiation/(MJ/m^2)
I	resource rich I	≥6700
II	resource rich II	5400~6700
III	resource rich III	4200~5400
IV	resource starvation IV	<4200

Solar Energy

The regional distribution of solar radiation in China: the western region has the highest solar radiation, followed by the eastern region, while the northern region has higher solar radiation than that of the southern part, and the plateau region has higher solar radiation than the plains do.

Specialized Vocabulary

- solar energy distribution　太阳能分布
- solar energy resource(s)　太阳能资源
- solar energy resource abundance　太阳能资源丰度
- total solar sunshine hours　太阳日照总时数
- northern hemisphere　北半球
- mid-latitude area　中纬度地区
- solar radiation　太阳辐射
- solar energy radiation quantity　太阳辐射量
- regional distribution　地域分布，区域分布
- plateau ['plætəu] area　高原地区

It is recorded that, humans have been using solar energy for more than 3000 years. However, the use of solar energy as a source of energy and power is only over 300 years old. The history of modern solar energy applications can be traced back to 1615, when French engineer Solomon de Cox invented the world's first solar-powered engine. The machine used solar energy to heat air and make it expand to do work and pump water. From 1615 to 1900, several solar powered devices were developed, which adopted concentrating methods to collect sunlight. The engine power was small, and the working medium was mainly water vapor, which was expensive and had little practical value. Most of the solar powered devices were researched and manufactured by individual solar enthusiasts. In 1839, French scientist Becquerel discovered the "photovoltaic (PV) effect", which is the physical basis of the PV technology. Subsequently, during the 20th century, solar energy technology gradually developed. In 1954, the Bell Labs developed a practical monocrystalline silicon PV cell with a photoelectric conversion efficiency of 6%, which set the foundation for the large scale application of the PV power generation. In the early 1970s, there was an upsurge in the application of solar energy around the world, which also had a huge impact on China.

At present, China's solar energy industry scale ranks first in the world, China has the largest production and usage of solar water heaters in the world, China is a major producer of solar cells. There are two relatively mature solar products in China, solar PV power generation system and solar water heating system. It is estimated that by 2030, China's installed solar power generation capacity will reach 350 million kW, accounting for 30% of the total; by 2050, the installed capacity will account for 52%, and the proportion of power generation will exceed that of thermal power generation, reaching 34%.

There are mainly the following ways to use solar energy.

1. *Solar energy light-thermal conversion*

Solar thermal devices receive and concentrate solar energy, and then convert it into thermal energy, which is further used for solar water heater, heating and refrigeration; solar drying of the agricultural and sideline products, medicinal material and wood; distillation, greenhouses and cooking; solar desalination; solar thermal power generation, to name but a few.

2. *Solar energy light-electric conversion*

This is direct photoelectric conversion (photovoltaic effects) and photothermoelectric/light-heat-electricity conversion. Direct photoelectric conversion is mainly that the solar cell directly converts solar energy into electric energy through the "photovoltaic effect" of the semiconductor material in it, and provides electric energy to the load in the form of a decentralized/distributed power system.

3. *Solar energy light-chemical conversion/photochemical conversion*

Under the irradiation of sunlight, substances undergo chemical and biological reactions under the action of solar photocatalysis or photoelectric catalysis, including photosynthesis, photolysis, and photosensitization reactions, which convert solar light energy into chemical energy and other forms of energy. It is stored in chemical bonds or breaks chemical bonds or stimulates active free radicals. The essence of the reaction is that the molecules and atoms in the substance absorb the energy of the solar photons and become "excited atoms". Some electronic energy states in the excited atoms change, which changes the valence bonds of some atoms. When the excited atoms return to the stable state, the photochemical reaction will occur. For example, solar energy is used to split water into hydrogen with a catalyst, but it is still in the research and development stage.

4. *Light-biological conversion*

It mainly refers to the photosynthesis of the green plants and microorganisms, which stores light energy in the organisms. Green plants use light energy to synthesize organic matter and oxygen from carbon dioxide and water in the air.

5. *Solar lighting* (light guide)

Sunlight or the solar cell device is used for indoor lighting. In addition, optical fiber can be used to introduce sunlight into dark places such as basements to solve the lighting problem in areas with poor sunlight. Solar lighting is mainly used in building corridor lighting, city lighting and other aspects.

The utilization of solar energy is being promoted more and more, and it has become the safest, greenest and most ideal way of generating electricity that people can take in modern society. Once solar energy is employed on a large scale around the world, environmental pollution caused by fossil energy can be reduced; the environment will be greatly improved, and some new solar energy application fields may be opened up.

> **Specialized Vocabulary**
> - installed capacity, installation 安装量，装机容量
> - solar water heater 太阳能热水器
> - solar drying/solar energy drying 太阳能干燥
> - solar desalination [diːsælɪ'neɪʃən] 太阳能海水淡化
> - solar thermal power generation 太阳能热发电
> - solar photocatalysis 太阳能光催化
> - photoelectric catalysis 光电催化
> - photosynthesis 光合作用
> - photolysis 光解作用
> - photosensitization 光敏作用
> - photovoltaic effects 光伏效应

Part Ⅰ Photovoltaic

Chapter 1
Photovoltaic overview

Photovoltaics comprises principally the technology that generates direct current electrical power from semiconductor under illumination. As long as light in the solar spectrum is shining on the solar cell, it generates electrical power. When the light stops or becomes too dull, the electricity stops. Solar cells never need recharging as a storage battery does. Some have been in continuous outdoor operation on the earth or in the space for over 30 years.

1.1 Photovoltaic (PV) history

In 1839, French physicist A. E. Henri Becquerel put two pieces of metal into an ionic solution and made a voltaic battery which would produce extra voltaic (electric) potential when exposed to light (PV effect). In 1873, British scientist Wilough B. Smith noticed that the light sensitivity of the selenium (Se) material, and deduced that the conductive ability of selenium increases proportionally with the amount of light it receives when the selenium material is exposed to light. In 1883, Charles Fritts developed the first functional, intentionally constructed photovoltaic cells on the basis of selenium. The selenium has been melted into a thin sheet on a metal substrate and pressed an Au-leaf film on as the top contact, which was around 30 cm^2 in area. He mentioned, "the current, if not wanted immediately, can be either stored where produced, in storage batteries, or transmitted a distance and used there". Later, people call those devices that have this effect "photovoltaic devices". Among them, the semiconductor p-n junction devices have the highest solar-power conversion efficiency, and these devices are often called "solar cells".

In the early 1950s, the researchers at the Bell Labs of the United States accidentally found that silicon is sensitive to light after the impurity processing when they were looking for a reliable power supply for a remote communication system, and they discovered that the p-n junction diodes could generate a stable voltage. Within a year, they had produced a monocrystalline silicon p-n junction solar cell with the efficiency of 6%. From 1961 to 1971, the research was focused on improving radiation-resistance ability and cost reduction. From 1972 to 1976, all kinds of monocrystalline silicon photovoltaic cells for space use were developed. Ultra-thin monocrystalline silicon photovoltaic cells were developed in mid 1970s. In 1976, the Nobel Prize for physics was awarded to Professor Mott for discovering the electronic process of the amorphous solidification. Then amorphous silicon photovoltaic cells came into being based on monocrystalline silicon and polycrystalline silicon.

1.2 Status of photovoltaics

According to the prediction of European Union (EU), solar power will account for more than 10% of the total energy consumption while renewable energy will claim 30% of the total energy supply by 2030. In addition, solar power will occupy more than 20% of the total energy consumption by 2050, renewable energy will claim 50% of the total energy supply.

Figure 1.1 is a diagram of all different energy types. It is obvious that PV plays a major role in the energy mix, especially from 2030, and it will become the dominant energy form near the end of this century. There are many factors that will influence this data but the general development trend is similar. Therefore, solar power will be applied on a large scale in about 2015 and especially after 2030, and it could be the dominant energy in all likelihood after 2050. However, the net cost of PV needs to be low enough and comparable to that from the existing grid in order to keep energy affordable.

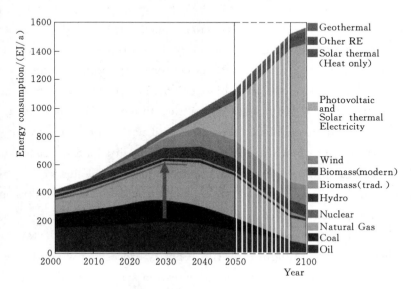

Figure 1.1 Scenario of the future energy supply in the world

1.2 Status of photovoltaics

> **Specialized Vocabulary**
> - PV power generation 光伏发电
> - solar power 太阳能发电
> - total energy consumption 总能耗、能源消费总量
> - total energy structure 总能源结构
> - geothermal 地热
> - solar thermal 光热，太阳能热，太阳热
> - dominant energy 主导能源
> - grid 输电网
> - State Grid 国家电网有限公司
> - energy mix 能源组成
> - natural gas 天然气
> - coal 煤
> - oil, petroleum 油，石油

The global photovoltaic market has broad prospects. According to the International Renewable Energy Agency (IRENA), the cumulative installed capacity of photovoltaics worldwide will reach 14,000 GW by 2050. The cumulative installation capacity of global PV in 2020 was 725 GW. From this, it is estimated that the compound growth rate between 2020 and 2050 will be 10.4%. There is still plenty of room for photovoltaic's development.

So far, the photovoltaic industry development has gone through four stages, which are shown in Figure 1.2.

(1) The initial stage of development during 2000 to 2010: The compound growth rate of the installed capacity reached 38.7%, mainly in Europe.

Since 2000, the global solar photovoltaic industry has entered a period of rapid development, and the annual installed capacity has increased rapidly. Germany enacted the *Renewable Energy Law* in 2000, which set a solid legal foundation for the fast development of the German PV industry. In 2004, Germany made the first amendment to the *Renewable Energy Law*, and this greatly increased the benchmark price of the PV power stations. The sudden increase in yields led to a large influx of capital, which led to the fast development of the German PV industry. Since then, Spain and Italy successively passed bills to subsidize solar photovoltaic power generation. So far, solar energy has been used more globally as a kind of clean energy.

Since the second half of 2008, the growth rate of the solar photovoltaic demand has declined due to the impact of the global financial crisis.

Since the second half of 2009, the demand of the photovoltaic market has regained the momentum of rapid growth, and China has also set off an investment boom in the photovoltaic industry.

Chapter 1 Photovoltaic overview

According to the "Global Market Outlook for Photovoltaics Until 2015" released by the European Photovoltaic Industry Association in May 2011, the cumulative installed capacity of global photovoltaics was 1.5GW in 2000, and increased by 39.5GW in 2010, with an average annual compound growth rate of 38.7%.

(2) The transitional period during 2011 to 2013: China replaced Europe and gradually became the world's largest photovoltaic market.

European governments have drastically cut photovoltaic subsidies, and European photovoltaic demand has shrunk rapidly, which has led to a slowdown in the growth rate of the global photovoltaic installations, and the photovoltaic industry has fallen into a trough. China was caught in a serious period of overcapacity dilemma. The United States and Europe successively launched "antidumping" and "countervailing" investigations on China's PV industry in 2011 and 2012, which caused a severe blow to the overall PV industry. The global PV newly-installed capacity fell for the first time in 2012. Japan and China successively issued supporting policies for the photovoltaic industry in 2013, and China's photovoltaic industry began to pick up. Since 2013, China, Japan and the United States have replaced Europe as the main growth regions for global photovoltaic installations.

(3) The growth period from 2014 to 2018: With the introduction of the photovoltaic subsidy policies by various countries across the world, photovoltaic development has entered the growth period, gradually shifting from being policy-driven to market-driven. China has successively introduced a series of policies to increase the share of new energy in the energy consumption.

(4) The parity period from 2019 to 2025: the main development is in the world.

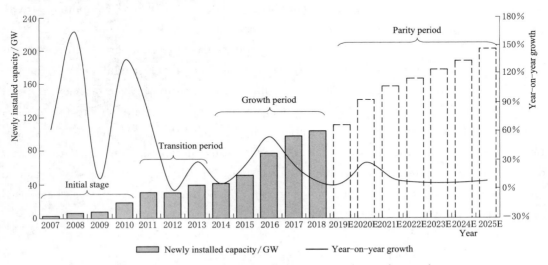

Figure 1.2 Global photovoltaic installed capacity and growth over the years

Due to the continuous advancement of technology, the cost of photovoltaic investment and power generation has decreased year by year, and the era of the PV parity has arrived. According to the statistics from Bloomberg New Energy Finance (Bloomberg-NEF), the global standardised electricity cost for PV in 2020 was 50 US/(MW·h), down 4% from 2019 and 80% from a decade ago. Photovoltaic power generation entered the era of parity in 2021. Figure 1.3 shows the decreasing trend of the global and China's photovoltaic power generation cost.

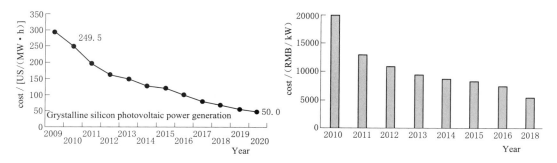

Figure 1.3 The decreasing trend of the global and China's PV power generation cost

Although it started late, China has developed fastly in photovoltaic power generation (Figure 1.4) and has now become a global leader in the development of the new energy.

In 2009, China launched a demonstration project for photovoltaic building applications. From 2011 to 2019, both the compound growth rate of the cumulative installed capacity of China's new energy power generation and the compound growth rate of power generation exceeded 30%. By the end of 2019, China's photovoltaic installed capacity was 205 GW, occupying for 33% of the global installed capacity.

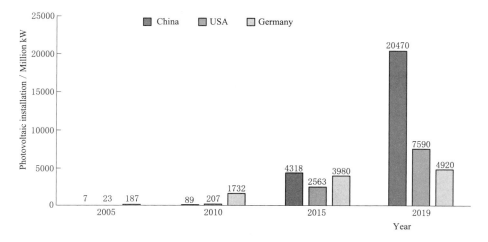

Figure 1.4 Photovoltaic installation in China, USA, and Germany

In addition, from the data of the newly installed photovoltaic capacity in China (Figure 1.5), reached 39.4 GW/a during 2016—2020.

Chapter 1   Photovoltaic overview

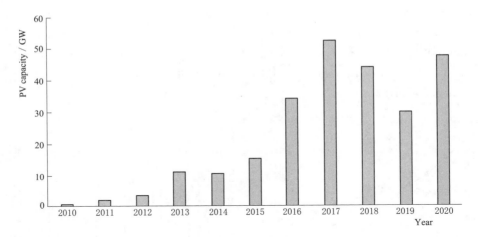

Figure 1.5   Newly installed PV capacity in China

> **Specialized Vocabulary**
> - International Renewable Energy Agency (IRENA)   国际可再生能源署
> - installed capacity   装机量，安装量
> - antidumping, countervailing   反倾销，反补贴（双反）
> - photovoltaic subsidy policies   光伏补贴政策

1.3   Applications of photovoltaics

1.3.1   Space application

The American satellite Vanguard-I was launched into orbit in 1958 with solar cells as its power system for signaling, which means that the new era of using solar cell for space power supply was opened up.

1. *Hubble Space Telescope*

It has two solar panels on each side. The panels are 11.8 m long, 2.3 m wide and produce 2.4 kW power (Figure 1.6).

2. *China's manned spacecraft - Shenzhou Ⅶ*

In Figure 1.7, the major power supply for Shenzhou Ⅶ is from its solar cell arrays. Its two wings are composed of the panel components, with 4 sections of solar panels for each wing, re-

Figure 1.6   Hubble Space Telescope with solar cell arrays

spectively. Solar arrays consist of a battery of monocrystalline silicon boron back surface field. The actual solar-power conversion efficiency is as high as 15%. The whole spacecraft uses 11,690 interconnected solar cells, and the cloth coefficient reaches 92.5%. Its solar-power conversion efficiency and cloth coefficient have met the international advanced level for similar products.

Figure 1.7 China's Shenzhou Ⅶ manned spacecraft with solar cell arrays

Subsequently, China has launched Shenzhou Ⅺ at Jiuquan Satellite Launch Center in October 2016, for which the main power supply is from GaAs solar panels.

> **Specialized Vocabulary**
> - space application　空间应用
> - space power　空间电源
> - grid-connected generation　并网发电
> - solar panel　太阳能电池板
> - efficient monomer battery　高效单体电池
> - monocrystalline silicon boron back surface field　单晶硅硼背场
> - solar-power conversion efficiency　光电转换效率，太阳能转换效率

1.3.2 Ground application

Ground applications include lighting, communications, transportation, Building Integrated/Attached Photovoltaic, and grid-connected (solar farms), which are the trends for PV technology application.

1. *Solar Lighting*

Solar lighting takes solar energy as its power; as long as the sunshine is sufficient, the cells can be installed locally with a rechargeable battery, without the involvement of the power supply grid. With no requirement of digging ditches or burying lines and no conventional power consumption, it is a kind of green environmental protection product.

Solar Street Lighting

Figure 1.8 shows a solar street light. It is comprised of solar modules, batteries,

Chapter 1 Photovoltaic overview

charge controller, lighting circuit, poles, sensors and other control components.

The pole is the supporting part of the whole system, which is different from the conventional street light. It needs to support not only the light holder, but also the solar modules and batteries, which are often near the base. Therefore, solar street lights at present preferably use the efficient crystalline silicon solar cells, and a choice of lamps of low power - consumption and high brightness (e. g. LED). The controller of the solar street lights should be able to be switched on and off automatically in addition to being equipped with such functions as to prevent reverse charging, overcharge, over discharge, short circuit and reverse lamp, which are possessed by common photovoltaic systems.

Figure 1.8 Solar street lights

Solar Garden Light (Figure 1.9)

The parameters of a solar garden light are listed below:

Solar modules: 2W

Battery: preferably Ni/Cd but Li - ion and Pb - ion are also used.

Light Source: LED

Lamp Body Material: Al alloy

Service Time: >8h

Continuous rainy days: 2~3 days reserved

The lamp body height: ~0.6m

Solar traffic warning light

The solar traffic warning light (Figure 1.9) may confront wind, hail, or ultraviolet irradiation, and must be unaffected by the environment, and sustain light for more than 10 days even over consecutive raining days.

2. *Solar bag, solar computer bag*

The built - in multifunctional solar emergency charger (Figure 1.10) may supply power by converting solar energy into electricity. The solar panels absorb solar energy at any time and anywhere, and the power can be stored in a built - in battery. This product is

1.3 Applications of photovoltaics

Figure 1.9 Solar garden light and solar traffic warning light

suitable for emergency use of power when you are in need of the charging of cell phones, digital cameras and other digital products outdoors.

The components of a typical solar bag are described as solar panels (waterproof), storage battery, charging adapter, connecting wiring, USB charging circuit, adapter and computer accessories, the configuration of which can be according to the customer specification.

3. *Solar car and yacht Solar car*

The solar electric car described herein is from the Institute of Solar Energy Systems, SunYat-sen University (though there are now many demonstration solar

Figure 1.10 Solar bags

cars). Its appearance is like a battery powered car in the park, capable of holding six passengers, but the highest speed is only 48 km/h, with travel time as short as one hour, shown in Figure 1.11. Of course, there are many recent advances.

Solar yacht

The first solar yacht made in China came into being in Zhuhai in July 2008. It was also one of the first solar yachts launched into market in the world, and before this most of the foreign solar yachts and boats were used for scientific experiment purposes or tourism promotion. Although in the aspect of speed, it may not yet be at a satisfactory level, the solar yacht made itself the star at the international boat show in Miami due to its "green" and

Chapter 1 Photovoltaic overview

Figure 1.11 Solar electric cars

environmental protection characteristics.

The yacht's sails are the key parts whose purpose is to concentrate direct and indirect sunlight, and they can also be used as ordinary sails. Controlled by the computer, the array can rotate around two axes, and its angle can be altered for better light concentration. The energy collected by solar modules is stored in the lead-acid batteries. Solar yacht is shown in Figure 1.12.

Figure 1.12 Solar yacht

4. *Building Attached Photovoltaics* (*BAPV*), *Building Integrated Photovoltaics* (*BIPV*)

Solar cells become part of the building itself in the application of BAPV and BIPV, such as roof tiles, curtain walls, doors, windows, and awning, to name but a few. There are many examples for BIPV and BAPV including those shown below in Figure 1.13.

1.3 Applications of photovoltaics

(a) Yingli PV research and development center

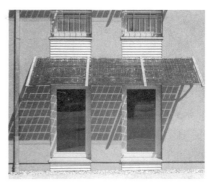

(b) Awnings: cover sunshine, keep off rain

(c) Photovoltaic curtain wall

(d) PV corridor

Figure 1.13 Four examples for BIPV and BAPV

Specialized Vocabulary

- ground application 地面应用
- solar light 太阳能灯
- solar street light 太阳能路灯
- solar modules 太阳电池组件
- battery 蓄电池
- charge controller 充放电控制器
- lighting circuit 照明电路
- pole 灯杆
- efficient crystalline silicon solar cell 高效晶体硅太阳电池
- low power‐consumption 低功耗
- high brightness 高亮度

Chapter 1 Photovoltaic overview

- lighting lamp　照明灯管
- reverse charging　反充
- overcharge　过充
- over discharge　过放
- switch on/off　开/关
- solar garden light　太阳能草坪灯
- solar traffic warning lights　太阳能交通警示灯
- ultraviolet irradiation　紫外线照射
- sustainable light　可持续发光
- built–in　内置
- multifunctional solar emergency charger　多功能太阳能应急充电器
- emergency charger　应急充电器
- converting solar energy into electricity　太阳能转换成电能
- absorb　吸收
- built–in battery　内置蓄电池
- battery (powered) car　电瓶车
- solar electric car　太阳能电动车
- solar yacht　太阳能游艇
- launched into market　投放市场，投入市场
- environmental protection　环境保护
- sail　帆
- lead–acid battery　铅酸蓄电池
- Building Attached Photovoltaics (BAPV)　光伏与建筑结合/一体化
- Building Integrated Photovoltaics (BIPV)　光伏建筑一体化
- photovoltaic curtain wall　光伏幕墙
- PV corridor　光伏走廊

5. *The application of photovoltaic power generation in Shanghai World Expo*

Solar power generation projects of Shanghai World Expo are examples of BIPV on the largest scale and of the highest technology in China. Solar power generation technology was applied more widely at the Shanghai World Expo than in the history of all the world expos. Chinese Pavilion with PV technology is shown in Figure 1.14.

The solar power generation technology was applied to the whole World Expo park, world expo center, Chinese Pavilion and Theme Pavilions. Among all these, just in the expo center, there was a large solar project on its roof with the capacity of 1 MW, comprising a large area of efficient solar modules, and on its walls were installed double–faced transparent sunshade solar modules. The array covers an area of 8000 m² and is able to generate electricity of 1 million kWh, and cut carbon dioxide emissions annually by 900

1.3 Applications of photovoltaics

Figure 1.14 Chinese Pavilion with PV technology

tons. Theme Pavilions with PV technology is shown in Figure 1.15.

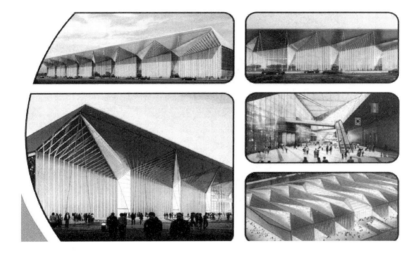

Figure 1.15 Theme Pavilions with PV technology

A "seven-colored flower" shown in the Jiangsu Pavilion in 2010 Shanghai World Expo, demonstrated to people how solar energy is captured across the light spectrum with the seven colors of red, orange, yellow, green, indigo, blue and purple reflected. This flower is in fact a solar device in the form of a sunflower made of solar panels. It can rotate around the sun, collecting seven colors of spectrums to gather new energy. Solar energy absorbed is transmitted to the daylighting import & storage system, where it can be converted into power (Figure 1.16).

The Japanese Pavilion was a building known as "Purple silkworm island", the outside of which is made of semi-transparent films that can absorb solar power and shine at night. The curved dome has three holes and three different angles. The holes are to receive rain for recycling use, and introduce sunlight for illumination to reduce the electricity demand. The angles are to enhance the circulation of warm and cold air and to reduce air conditioning energy consumption (Figure 1.17).

Chapter 1 Photovoltaic overview

Figure 1.16 Jiangsu Pavilion with PV technology

Figure 1.17 Japanese Pavilion with PV technology

Specialized Vocabulary

- solar (power generation) projects 太阳能发电项目
- capacity 容量
- MW, megawatt 兆瓦，百万瓦特
- efficient solar modules 高效太阳电池组件
- solar devices 太阳能装置
- solar panels 太阳电池板
- daylighting import system 采光导入系统
- convert into power 转化为电力
- energy consumption 能耗
- generate electricity 发电
- cut carbon dioxide emissions 二氧化碳减排

6. *Photovoltaic power station* (*solar farm*)

Areas with no electric power in the west of China will depend to a large extent on the photovoltaic power station to provide electricity.

By 2012, the total photovoltaic power station capacity of Qinghai had grown to over 500 million kWh, with the installed capacity of photovoltaic power stations being 7% of the provincial total.

At present, there are 42 grid-connected photovoltaic power stations in Qaidam basin in Qinghai province, with the capacity of 1003 MW. Qinghai has made several achievements as follows, creating a photovoltaic power station with the largest installed capacity within the region, the largest grid-connected photovoltaic power station in the world at the date of installation, and the first grid-connected PV power station of mega watt capacity in the world. A PV power station in Qinghai Province is shown in Figure 1.18.

Figure 1.18 A PV power station in Qinghai Province

> **Specialized Vocabulary**
> - photovoltaic power station 光伏电站
> - grid-connected 并网（发电）
> - installed capacity 装机/安装容量
> - solar energy utilization patterns 太阳能利用方式
> - the total capacity 总发电量

1.4 Technology trends

The trends of the solar cell technology in the future will include ①flexible, thin-film solar cells with the characteristic of low cost, high efficiency and long service life, ②Multi-junction tandem cells such as GaInP/GaAs/New/Ge for which new materials are the challenge, and ③concentrator solar cells. Future development goals are to improve the

Chapter 1 Photovoltaic overview

efficiency, reduce manufacturing costs and expand the application field.

The second technology trend is to develop the grid – connected PV power plant with extra large capacity (>100 MW and GW level) plus grid – connected PV power plant integration technology, and large – scale grid – connected inverter and efficient cost effective concentrator PV power generation system.

The third technology trend is BIPV. In future, people will expect to use new photovoltaic building materials and components to access integrated PV easily, with a standard photovoltaic price.

Reading Materials

1. *A review of photovoltaic poverty alleviation projects in China: Current status, challenge and policy recommendations*

2. *Review on photovoltaic power and solar Resource forecasting: current status and trends*

Reading Materials

Exercises and Discussion

1. What do you think of the solar PV situation in China?

2. Please describe the mechanism of the photovoltaic effect.

3. What do you think of the solar PV application relevant to the present market? And explain why it will be the major new energy around 2050.

4. What are the utilization patterns for solar energy?

Chapter 2
Principle of photovoltaic

Photovoltaic power generation is a major technological revolution in the history of human power generation. It is not only different from various chemical batteries that convert chemical energy into electricity, but also different from traditional generators that convert thermal energy or mechanical energy into electrical energy based on the classical electrodynamic principles. The photovoltaic conversion process is a pure physical process that converts photon energy into electron energy by the PV effect of the photovoltaic materials. The whole process itself does not consume any other resources and substances except light. In addition to releasing a certain amount of heat, it does not emit any other chemicals while generating electricity. It is an "absolutely clean" power generation process.

2.1 P – n Junction of the solar cells

Photovoltaic materials used in the photovoltaic power generation are an important class of semiconductor materials. Currently commonly used semiconductor materials include Si, Ge, GaAs, and CdTe. These semiconductor materials have the unique photovoltaic conversion properties. The conduction of the semiconductor materials is achieved by the directional movement of two kinds of carriers, namely electrons and holes.

Silicon is an important semiconductor material. Its atomic structure brings it to be an ideal elements in photovoltaic solar cells, namely, with fourteen electrons and only 4 electrons in the outermost electron shell. Generally, the shell needs 8 electrons to be stable. Therefore, in order to obtain a stable silicon crystal, each silicon atom will combine with four other silicon atoms in its normal or pure state. In pure state of silicon crystal, the free electrons to carry the electric current is very few. Therefore, to change their electrical conductivity, other elements as dopants are doped into silicon. Moreover, silicon is non-toxic, abundant in the world, relatively cheap and a mature commodity.

The indirect band gap of crystalline silicon is around 1.17 eV, the direct band gap is above 3 eV at the ambient temperature, which determine the optical properties' variation of Si with wavelength, including the low absorption coefficient to generate carrier of the near band gap photons. At short ultraviolet wavelengths, two electron – hole pairs' generation by one photon seems possible, though quantitatively this is a small effect; at the other extreme of the spectrum parasitic free – carrier absorption competes with band – to – band generation. The intrinsic concentration is another important parameter related to the band structure, which links carrier disequilibrium with voltage.

The p – n junction formed after doping in silicon is the core of most semiconductor devices, the main structural unit of the solar cells, and the basis of the photovoltaic conversion of the solar cells. The p – n junction results from "doping" process that generates valence – band or conduction – band selective contacts with one becoming the n – side (a great many negative charge), the other the p – side (a great many positive charge), that is, the p – n junction is generated by connecting p – type and n – type semiconductor materials (Figure 2.1).

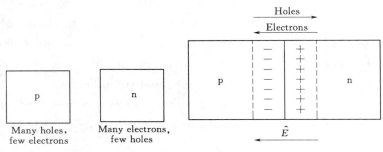

Figure 2.1 The formation of a p – n junction[1]

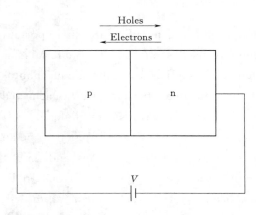

Figure 2.2 Space charge region in p – n junction[2]

After joining p and n type materials together, the excess holes in p type area diffuse into n type area, while electrons in n type area diffuse into p type area as a result of the carrier concentration gradients across the junction. The electrons and holes leave behind exposed charges on dopant atom sites, fixed in the crystal lattice. Thus, the electric field ($\hat{E}$) is built up in the depletion region of the junction to inhibit diffusion. With the materials used, the 'built – in' potential (V_{bi}) owing to $\hat{E}$ will be formed. If a voltage is applied to the junction (Figure 2.2), $\hat{E}$ will be reduced. In other

---

[1] Stuart R. Wenham, Martin A. Green, Muriel E. Watt, Richard Corkish, Alistair Sproul. Applied Photovoltaics [M]. Abingdon: Earthscan, 2011.

[2] Stuart R. Wenham, Martin A. Green, Muriel E. Watt, Richard Corkish, Alistair Sproul. Applied Photovoltaics [M]. Abingdon: Earthscan, 2011.

words, when the solar cell absorbs sunlight, the resulting non-equilibrium carriers enter the built-in electric field, also will drift toward two ends of the space charge region, respectively, thereby generating a photoelectric potential (voltage), which is the PV effect. If the p-n junction is connected to the external circuit, current will appear in the circuit, which is the basic principle of the photovoltaic conversion of the solar cells.

> **Specialized Vocabulary**
> - p-n junction　p-n结
> - doping　（在半导体材料中）掺杂
> - conduction-band　导带，传导带，导电带
> - valence-band　价带
> - negative charge　负电荷
> - positive charge　正电荷
> - carrier　载流子
> - depletion region　耗尽区，势垒区，阻挡层
> - built-in potential　内建电势，内建电场，内建电位
> - electric field　电场

2.2　Solar cell types

Since the invention of the silicon solar cells in the 1950s, research and industrialization of the solar cells have been greatly developed. New processes, materials and structures of the solar cells emerge one after another. We can divide solar cells into silicon-based solar cells, inorganic compound solar cells and new organic semiconductor thin film solar cells according to the materials used in them. Among them, silicon-based solar cells with very mature technology have been developed for the longest time, which include monocrystalline silicon, polycrystalline silicon and amorphous silicon solar cells. Inorganic compounds thin-film solar cells include Cadmium Telluride (CdTe), Copper Indium Selenide (CIS), Copper Indium Gallium Selenide (CIGS) and Gallium Arsenide (GaAs). The new concept solar cells include Organic solar cells/Organic photovoltaics (OPV), dye-sensitized solar cells (DSSC), perovskite solar cells (PSC), and quantum dots solar cells (QD-SC). The big difference of the last type of solar cells from the others is that some introduce organics and nanotechnology into the production process without the minority carriers, and develop the PV effect through the kinetics of the charge transfer.

There are different ways of classifying the existing solar cells.

(1) According to the base material, there are four types of solar cells.

1) Crystalline silicon solar cells, including Monocrystalline silicon and Polycrystalline silicon.

Chapter 2 Principle of photovoltaic

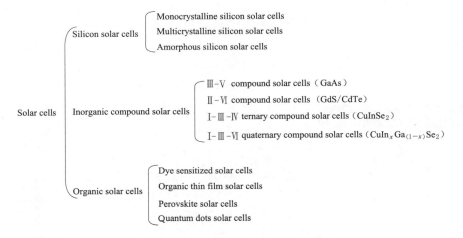

Figure 2.3 Solar cell classification

2) Amorphous silicon solar cells.

3) Inorganic compound solar cells, including GaAs, CdTe, and CIGS.

4) Organic & nanotechnology semiconductor solar cells, including DSSC, OPV, PSC and QDSC.

(2) According to the structure of the solar cells, there are five types.

1) Homojunction solar cells, including silicon solar cells and GaAs cells.

2) Heterojunction solar cells, such as SnO_2/Si and $Si/GaAs$.

3) Schottky junction solar cells, such as Metal – Semiconductor.

4) Composite/Multi – junction solar cells, including Multiple p – n junctions.

5) Liquid junction solar cells, such as photoelectrochemical solar cell.

(3) According to the application of the solar cells, there are solar cells applied in space, Terrestrial solar cells, and Light sensors.

(4) According to the Work Mode, there are flat solar cells and concentrator solar cells.

Maximum Efficiencies for Solar Cell Materials

The best efficiencies for all solar cells are shown in Figure 2.4 based on the data from National Renewable Energy Laboratory (NREL) until 2022. The highest efficiency is normally from the concentrated solar cells with four or more junctions, which is 47.1%. For silicon – based cells, best efficiency is over 25%, which is from the monocrystalline silicon solar cells (non – concentrator). The highest efficiency for the compound solar cells is from single – junction GaAs with 27.8%. For the new generation solar cells, the highest efficiency is obtained by the perovskite solar cell which is the latest developed cell and the fastest growing solar cell to date.

2.2 Solar cell types

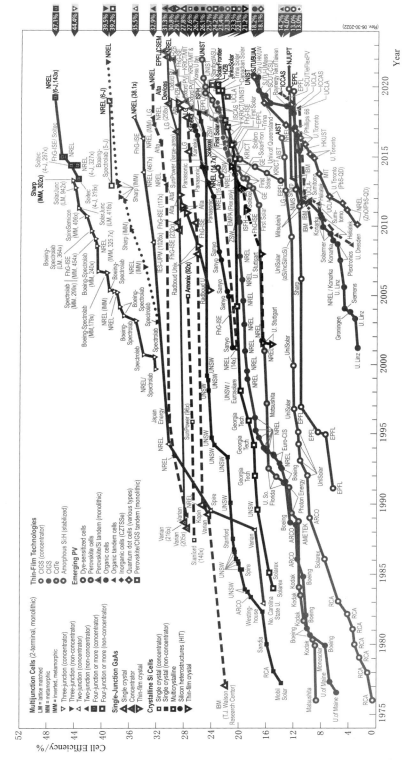

Figure 2.4 Best research-cell efficiencies from NREL

Chapter 2 Principle of photovoltaic

> **Specialized Vocabulary**
> - monocrystalline silicon 单晶硅
> - polycrystalline silicon 多晶硅
> - amorphous silicon 非晶硅
> - inorganic compounds thin-film solar cells 无机化合物薄膜太阳电池
> - organic solar cells/Organic photovoltaics (OPV) 有机太阳电池
> - dye-sensitized solar cells (DSSC) 染料敏化太阳电池
> - perovskite solar cells (PSC) 钙钛矿太阳电池
> - quantum dots solar cells (QDSC) 量子点太阳电池
> - light sensors 光感应器
> - flat solar cells 平板太阳电池
> - concentrator solar cells 聚光太阳电池

2.3 The structure and working mechanism of the solar cells

Figure 2.5 shows the appearance and the basic structure of the silicon solar cell. The basic material is the p type monocrystalline silicon with the upper surface which is n+ type area, together with p-n junction. There is metal palisade electrode (structure to allow light to enter the cell) on the top surface, and the metal bottom electrode is on the back of the silicon. The top and bottom electrodes contact with n area and p area respectively to form ohmic contacts. There is also the antireflection film covered evenly on the whole top surface.

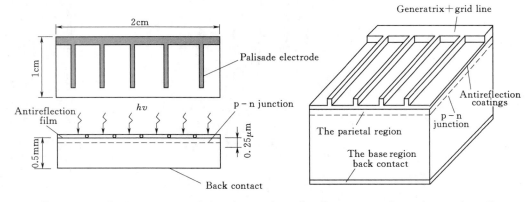

Figure 2.5 Basic structures of the silicon solar cell and monocrystalline silicon solar cell

The specific components of monocrystalline silicon solar cells are described as below.

(1) The back ohmic contact area. The metal and the base region (single crystal substrate) form ohmic contact at the back surface.

(2) The base region. The base region consists of the single crystal substrate materials. The

2.3 The structure and working mechanism of the solar cells

function of the substrate is to support the solar cells. The substrate for the single crystalline solar cell is a silicon wafer, and for the thin film solar cells, it can be glass, metal foil or flexible materials (Figure 2.6).

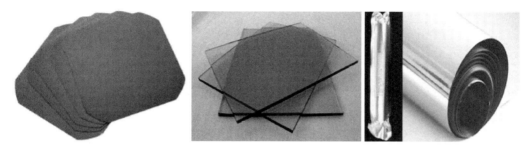

Figure 2.6 Different substrates: silicon wafer, glass, metal foil

(3) The junction area is namely, the p-n junction barrier area. The both sides of the junction can be made of the same material or the different materials. If the p-n junction is composed of the same semiconductor materials (only of the different conductive type), it is known as the homojunction. If the p-n junction is made of different materials, it is known as the heterojunction. Generally, the single crystal solar cells are using the homojunction, while the thin film solar cells developed more recently are using the heterojunction. In addition, in the single crystal and thin-film solar cells, the Schottky barrier area also has been used.

The p-n junction is used to generate the PV effect. The function of the p-n junction is absorption and conduction. In solar cells, the absorption of the photons occurs at the p-type layer. It is important to have a better absorption ratio and generate more conduction electrons in the p-type layer. Another important goal is reducing the recombination of the electrons and holes to extend the charge lifetime of the solar cells.

(4) The top region. Another layer of semiconductor forms the top region, which is in contrast to the material of the conductive area, where the top region is much thinner than the base region. In the case of the front illumination, most of the light (especially the long wavelength photon) crosses the top region and the junction area to reach the base region, and is absorbed within the thickness of the base region which is closed to the p-n junction, and then produces electron-hole pairs. Most of the minority carriers diffuse and drift to the parietal area to form the photocurrent, and the recombination of a few minority carriers happens in the diffusion process.

(5) The metal grid contact (including grid line and the generator). They covered about 5%~10% of the photosensitive area. The metal electrode in the solar cells is used as electric contact (Figure 2.7). There are back electrode and front electrodes. Generally, Aluminum (Al) or Molybdenum (Mo) are used as the back electrode. The front electrode is normally in a grating pattern in order to reduce the shading losses. Physical Vapor Deposition (PVD) and chemical vapor deposition (CVD) have been used to deposit the

Chapter 2 Principle of photovoltaic

Figure 2.7 Metal electrode in grating pattern

metal electrode to achieve good performance, but it has high cost.

(6) Antireflection coating (ARC). The purpose of the antireflection coating is to reduce the reflection loss of the incident light, increase the proportion transmitted and the photocurrent as well. The ARC materials commonly used are TiO_2, SiO_2, Ta_2O and ITO (indium tin oxide). For the silicon-based solar cells, silicon is shiny gray and can reflect up to 30% of the incident light, which will reduce the energy transfer efficiency. With a coating of the anti-reflection layer which has a texture surface structure, the incident light captured will go up to 10%~15% (Figure 2.8).

(7) Encapsulation. The commonly used ma-

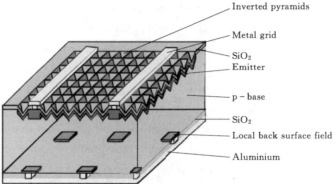

Figure 2.8 Surface textures as anti-reflection layer for solar cells

terials for encapsulation are glass and cellophane tapes.

(8) If it is the thin film solar cell, the above structure will all be deposited on a transparent glass or other cheap substrate.

There are some performance requirements of the materials used for the solar cells. The bandgap of the materials is between 1.1 eV and 1.7 eV, and it is preferably a direct bandgap material, non-toxic material, able to be produced in large areas, with high photovoltaic conversion efficiency and long term stability.

The power generation principle of the solar cell is based on the photoelectric effect caused by the light irradiating on the semiconductor. In irradiation, the photons' energy is larger than the forbidden band width which will cause the production of the electrons and the holes. The p-n junction in the semiconductor will allow the electrons to diffuse to the n-type semiconductor and the holes to diffuse to the p-type semiconductor, namely, the negative charges and the positive charges will gather at both sides, respectively. Therefore, when these two electrodes are connected, the charges will flow to generate electricity. This is completely different from the traditional power generation, in producing DC power, without the rotating parts from the motor or exhaust gas, so it is a clean, noiseless electric generator.

> **Specialized Vocabulary**
> - ohmic contact 欧姆接触
> - antireflection film 减反射膜
> - antireflection coating (ARC) 抗反射层
> - base region 基区、基极区
> - homojunction 同质结
> - heterojunction 异质结
> - the junction area 结区
> - the top region 顶区
> - electron-hole pairs 电子-空穴对
> - the parietal area 顶叶区
> - the metal grid contact 金属栅接触
> - grid line 栅线
> - generator 发电机、发生器、产生器
> - forbidden band 禁带
> - Physical Vapor Deposition (PVD) 物理气相沉积
> - Chemical Vapor Deposition (CVD) 化学气相沉积

2.4　The fundamental characteristics of the solar cells

The fundamental characteristics of the solar cells are the polarity, the performance parameters, and the volt-ampere characteristics, as follows.

(1) The polarity of the solar cell. The silicon solar cell is generally made in p+/n type structure or n+/p type structure. P+ and n+ mean the conduction type of the conductive materials is from the front illumination layer of the solar cell. N and p mean the conduction type of the substrate conductive materials from the back side of the solar cell. The solar cell's electrical properties are related to the characteristics of the semiconductor materials used for the solar cells.

(2) The performance parameters of the solar cell. These include open-circuit voltage, short-circuit current, maximum power output, fill factor and transfer efficiency. The parameters can evaluate the performance of the solar cell.

Chapter 2 Principle of photovoltaic

(3) The volt–ampere characteristic of the solar cell. Generally, the solar cell module, a multimeter and a load resistance are used to create the circuit, and the load resistance R is changed, measuring the current and the voltage drop across the load, then graphing the volt–ampere characteristic curves of the PV module.

Specialized Vocabulary

- polarity 极性
- volt–ampere characteristic 伏安特性
- open–circuit voltage 开路电压
- short–circuit current 短路电流
- maximum power output 最大输出功率
- fill factor 填充因子
- (electron) transfer efficiency 转换效率
- solar cell module 太阳电池组件
- load resistance 负载电阻
- multimeter 万用表

Reading Materials

1. *Device characteristics and material developments of indoor photovoltaic devices*
2. *The Photovoltaic (PV) Effect*

Reading Materials

Exercises and Discussion

1. What is the p–n junction?
2. What kind of characterizations that we can use for the solar cells?
3. Please describe briefly about different types of solar cells.
4. What are the structure and mechanism of the silicon solar cells?

Chapter 3
Solar cells

The device that can directly converts light energy into electrical energy via the photoelectric effect called solar cells. According to the previous classification of the solar cells, we will introduce several main solar cells in detail below.

3.1 Silicon–based solar cells

Silicon solar cells mainly include the crystalline silicon solar cells (monocrystalline silicon solar cells and polycrystalline silicon solar cells) and amorphous silicon solar cells.

The crystalline silicon solar cell is currently the solar cell with the most mature technology and the largest industrial application scale, having a very high cost performance. The crystalline silicon solar cell is made of the crystalline silicon wafer as the main raw material, on which p–n junction, solar light trapping layer, electrode and other structures are fabricated. Depending on the different silicon wafers, it can be divided into the monocrystalline silicon solar cells and the polycrystalline silicon solar cells. In terms of the solar cell structure, the common structures of the two solar cells are not significantly different. It is basically composed of the front electrode, surface anti–reflection layer, p–n junction, and back electrode. As shown in Figure 3.1, it is basically composed of the front electrode, surface anti–reflection layer, p–n junction, and back electrode.

Crystalline silicon solar cells and modules have led the way of the PV technology since its inception, constituting more than 80% of the present photovoltaic market. Despite repeated announcements that they will be replaced by other technologies, they will remain their dominant for some time, at least for the next decade. One of the reasons for the dominance of crystalline silicon in PV is that microelectronics tech-

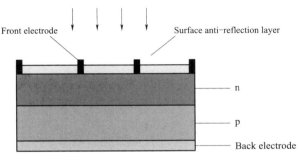

Figure 3.1 Common structure of the silicon–based solar cells

nology has greatly advanced silicon technology. On the one hand, the PV industry not only benefits from accumulated knowledge, but also has access to silicon feedstock and used equipment at reasonable prices. Microelectronics, on the other hand, takes advantage of some of the innovations and developments in the PV field.

Silicon and other semiconductor materials used for the solar cells can be monocrystalline, multicrystalline/polycrystalline, microcrystalline or amorphous silicon materials. Although the usage of these terms varies, there is a definition of planar grain size according to Basore (1994). The grain size of the microcrystalline material is less than 1 μm, polycrystalline less than 1 mm, and multicrystalline less than 10 cm. The structure of different material types is shown in Figure 3.2.

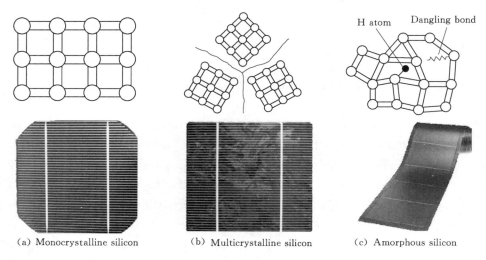

(a) Monocrystalline silicon   (b) Multicrystalline silicon   (c) Amorphous silicon

Figure 3.2　Different structures and surface morphology of silicon based materials and cells

As shown in Figure 3.2, the structure among the different types of crystalline silicon is different. For the monocrystalline silicon (c - Si) material [Figure 3.2 (a)], the atoms are arranged in a regular pattern with periodicity and are ordered. Multicrystalline or polycrystalline silicon material [Figure 3.2 (b)] is composed of single crystals of silicon with different crystal orientations, irregular shapes and grain boundaries, and the regions of the crystalline silicon are separated by the "grain boundaries", where bonding is irregular, but the atoms in each silicon crystal are arranged in a regular pattern. For the amorphous silicon material with a less regular arrangement [Figure 3.2 (c)], the atoms inside each "crystal" are disordered and unsystematic, which enables the material to bend freely and be "soft". This disorganized arrangement leads to the "dangling bonds" in the silicon that can be passivated by hydrogen.

It has been noted that in some cases, the characteristics of multicrystalline silicon cells may preclude the use of standard processing techniques. Some of the proposed alternatives are not cost - effective enough to be incorporated into industrial production lines,

3.1 Silicon-based solar cells

but others are already being utilized. Two main differences with monocrystalline silicon materials can be highlighted: ①The quality of polycrystalline materials is 'poorer' due to crystal defects (such as grain boundaries and dislocations) and metal impurities (dissolution or precipitation), thereby reducing the volume lifetime and thus the cell efficiency. To solve this problem, there are two main strategies: implementing the "gettering" step and passivation of the defect with hydrogen. ②Texturing is even difficult due to the different exposed crystal planes, so standard alkaline solutions are not suitable. To improve light-trapping and absorption, other techniques must be employed.

From Figure 3.2, the crystalline silicon solar cells derive their name from the way they are made. At the solar panel level, the difference between monocrystalline and polycrystalline is that the former is cut into thin wafers from a single continuous crystal grown for this purpose. Polycrystalline cells are made by melting the silicon material and pouring it into a mould. The uniformity of a mono-crystalline cell makes it dark blue all over. This also makes it more efficient than polycrystalline solar modules, which have a 'jumbled' surface that appears in various shades of blue. Except for the crystal growth stage, there is little difference between the structure of mono- and polycrystalline solar cells. The cells are usually laminated with tempered glass on the front and plastic on the back. These are connected by a transparent adhesive, and then the modules are framed with aluminum. The single crystal modules are smaller in size per watt than their polycrystalline counterparts.

Specialized Vocabulary
- band gap 带隙，能隙，能带间隙
- indirect band gap 间接带隙
- absorption coefficient 吸收系数
- intrinsic concentration 本征浓度
- grain boundaries 晶界

3.1.1 Monocrystalline silicon solar cells

As mentioned above, the monocrystalline silicon material has an ordered crystal structure, with each atom ideally lying at a predetermined position. As such, it allows for off-the-shelf applications of theories and techniques developed for crystalline materials and displays predictable and consistent behavior. However, it becomes the most expensive type of silicon because it requires a meticulous and slow manufacturing process. As a result, cheaper multicrystalline or polycrystalline silicon, and to a lesser extent amorphous silicon, are increasingly being used in solar cells, even though their quality is less than ideal.

Monocrystalline silicon cells are made of the cells saw-cut from a single cylindrical

Chapter 3 Solar cells

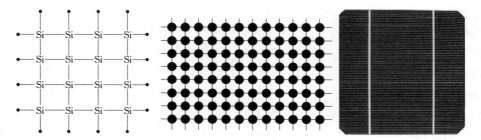

Figure 3.3 Crystal structure of the monocrystalline silicon material

crystal of silicon, which are effectively a slice from a crystal. The Czochralski technique (CZ), Float Zone (FZ) method and epitaxial method are used to produce the single crystalline silicon. The principle advantage of the monocrystalline cells is their high lab efficiency over 24%, although the manufacturing process required to produce monocrystalline silicon is complicated, resulting in slightly higher costs than other technologies. In appearance, it will have a smooth texture with the thickness of the slice visible. Being brittle, they must be mounted in a rigid frame and substrate to protect them.

The structure of the monocrystalline silicon solar cell is shown in Figure 3.4, which is mainly composed of front electrode, anti-reflecting coating, n-type silicon, p-type silicon and rear electrode.

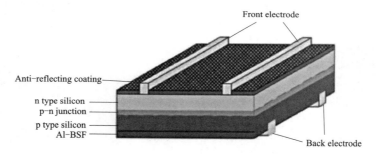

Figure 3.4 The structure of the monocrystalline silicon solar cell

Monocrystalline cells were first developed in 1955. They conduct and convert solar energy into electricity. When sunlight hits a silicon semiconductor, the light absorbs enough energy to loosen the electrons, allowing them to flow freely. Monocrystalline silicon solar cells are designed in such a way that free electrons can be directed in the cell's electric field, in a path or a circuit as currents, and then used to power various appliances. The power of a cell (in watts) is determined by a combination of the cell's current and voltage. The voltage depends on the electric field inside of the cell.

The disadvantages include the high temperature required to form the boule (ingot) of the pure silicon, the energy intensive manufacturing process, and the use of a relatively large amount of Si, which is expensive and fragile with a low band gap (1.17 eV $\approx$ 1060 nm).

3.1 Silicon-based solar cells

> **Specialized Vocabulary**
> - energy intensive manufacturing process 能源密集型制造过程
> - texturing 织构化
> - texture etch 织构刻蚀，表面刻蚀
> - plasma etch 等离子体刻蚀
> - sinter 烧结
> - diffusion 扩散
> - edge insulation processing 边缘绝缘处理
> - Czochralski Technique (CZ) 直拉法
> - Float Zone method (FZ) 区熔法
> - epitaxial method 外延法

3.1.2 Multicrystalline silicon solar cells

The technology to produce multicrystalline or polycrystalline silicon, is less critical and therefore cheaper than the technology needed to produce single-crystal materials. Grain boundaries degrade cell performance by blocking carrier flow, allowing the presence of additional energy levels in the forbidden gap, thus providing efficient recombination sites and a shunt path for current to flow through the p-n junction. To avoid significant recombination losses at the grain boundaries, grain sizes need to be at least a few millimeters. This also allows individual grains to extend from the front to the back of the cell, providing less resistance to carrier flow and generally reducing grain boundary length per unit cell. The polycrystalline material is widely used in commercial solar cell production.

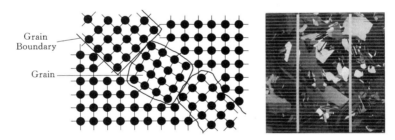

Figure 3.5 Crystal structure of the multicrystalline silicon material

Polycrystalline or multicrystalline silicon cells are made from cells cut from molten and recrystallized silicon ingots. In the manufacturing process, the molten silicon is cast into polycrystalline silicon ingots, which are then saw-cut into very thin wafers as single-crystal material and assembled into complete cells. Because of the simpler manufacturing process, multicrystalline cells are cheaper to produce than mono-crystal cells. However, they tend to be slightly less efficient, with the average efficiency of about 22%. They have a speckled crystal reflective appearance and again need to be mounted in a rigid frame, due to brittleness.

Chapter 3 Solar cells

Specialized Vocabulary

- shunting path　分流路径，分流电路
- recombination loss　复合损失
- grain size　晶粒尺寸
- ingot　铸块，锭
- molten silicon　熔融硅
- cast　浇铸
- speckled crystal reflective appearance　斑点晶体发光的外观
- rigid frame　刚架、刚性构架

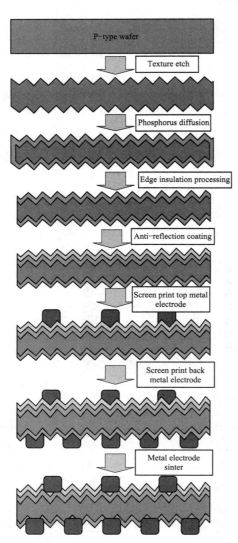

Figure 3.6 The basic solar cell manufacturing process diagram

The fabrication process of the monocrystalline silicon solar cells and the polycrystalline silicon solar cells is similar.

The basic manufacturing process diagram is shown in Figure 3.6.

(1) The silicon wafer is cleaned and the damaged layer is removed, and the surface of the silicon wafer is textured (roughened). The purpose of texturing on the surface of the silicon wafer is to reduce the reflection of the sunlight on the cell surface. Generally, random pyramid texture is prepared by alkaline etching solution for the monocrystalline silicon wafer, and random etching pit texture is prepared by acidic etching solution for the polycrystalline silicon wafer, but the obtained anti-reflection effect is not as good as that obtained by the alkaline etching of the monocrystalline silicon wafer.

(2) P-n junction prepared by the phosphorus diffusion. Phosphorus oxychloride is used as the phosphorus source, and the phosphorus atoms are diffused and doped into the surface of the p-type silicon wafer through a high-temperature diffusion process, so that the surface of the silicon wafer is inverted into an n-type, thereby forming a p-n junction.

(3) Edge insulation treatment. Edge p-n junctions are removed to prevent short circuits

on the front and back surfaces of the cell.

(4) Anti-reflection film coating. The silicon nitride antireflective layer deposited by plasma-enhanced chemical vapor deposition (PECVD), which can further reduce the reflection of the sunlight on the cell surface on the basis of the texture.

(5) Screen printed electrodes. In order to avoid electrode shading and ensure a small electrode resistance, the grid-shaped silver electrode is screen printed on the front surface of the cell; if the back surface of the cell does not require light entry, a full-surface back electrode can be used, usually printed with the cheaper Al electrode.

(6) High temperature sintered electrode. The silver electrode needs to be processed at 800~900℃ after the screen printing. Al is sintered at about 600℃ and diffuses into the silicon wafer to form an Al Back Surface Field (AlBSF). This can make the n-type back surface from the diffusion process become p-type through the inverse compensation doping.

The crystalline silicon solar cell prepared by the above process steps is simply called the aluminum back field crystalline silicon solar cell.

With the continuous progress of solar cell preparation technology, many new high-efficiency structures have been born in crystalline silicon solar cells. Since these high-efficiency structures can play a more synergistic effect on the monocrystalline silicon wafers, the application of polycrystalline silicon wafers in new high-efficiency crystalline silicon solar cells is becoming less and less. According to the continuous progress of crystal silicon solar cell structure technology, a variety of typical solar cell structures have appeared. ①Conventional aluminum back field crystalline silicon solar cells; ② PERC crystalline silicon solar cell; ③PERL crystalline silicon solar cell; ④PERT crystalline silicon solar cell; ⑤MWT/EWT crystalline silicon solar cells; ⑥IBC crystalline silicon solar cell; ⑦HIT crystalline silicon solar cell; ⑧HBC crystalline silicon solar cell; ⑨Black silicon solar cells. Figure 3.7 shows several of these cell structures.

Czochralski silicon (Cz-Si)

For decades, the terrestrial PV market has been dominated by p-type Czochralski silicon substrates. Continuous improvements in performance, yield, and reliability have led to substantial cost reductions and subsequent expansion of the PV market. Due to the low cost of multicrystalline silicon (MC-Si) wafers, MC-Si cells emerged in the 1980s as an alternative to monocrystalline silicon wafers. However, their lower quality prevents achieving similar efficiencies as Cz, so the figure of merit $/W for both techniques are very similar for a long time.

3.1.3 Amorphous silicon solar cell

Amorphous Silicon (a-Si) solar cells were developed in 1976 and commercialized in 1980, and currently have a maximum laboratory conversion efficiency of 14%. Amorphous silicon solar cells are composed of silicon atoms in a thin, uniform layer rather than a

Chapter 3 Solar cells

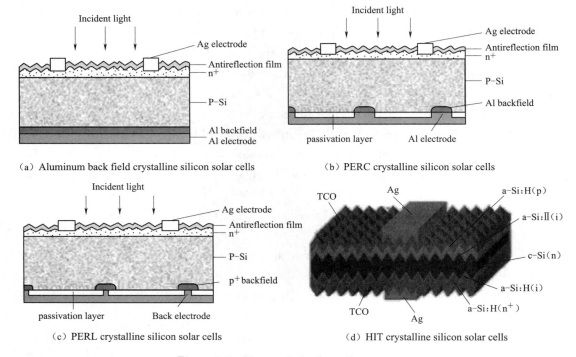

Figure 3.7 New typical solar cell structures

crystal structure. Amorphous silicon absorbs light more efficiently than crystalline silicon, so the cells can be thinner. Therefore, amorphous silicon is also known as "thin film" photovoltaic technology.

In principle, amorphous silicon is even cheaper to produce than polycrystalline silicon. For amorphous silicon, the atomic structures are arranged in no long-range order, resulting in certain regions within the material containing unsatisfied bonds, or 'dangling' bonds. This in turn leads to the creation of additional energy levels within forbidden gaps, making it impossible for pure semiconductors to be doped, or to obtain reasonable current flows in solar cell configurations. It has been found that the addition of hydrogen atoms to the amorphous silicon at the level of 5% ~ 10% can saturate the dangling bonds and improve the quality of the material. It also increases the band gap (Eg) of crystalline silicon from 1.1 eV to 1.7 eV, making the material more strongly absorb photons with energies above the latter threshold. Thus, the material thickness required to form an effective solar cell is much smaller. It has been found that the addition of hydrogen atoms to the amorphous silicon at the level of 5% ~ 10% can saturate the dangling bonds and improve the quality of the material as shown in Figure 3.8.

Amorphous silicon can be deposited on a variety of rigid and flexible substrates, making it ideal for curved surfaces and "foldable" modules. However, amorphous cells are

3.1 Silicon-based solar cells

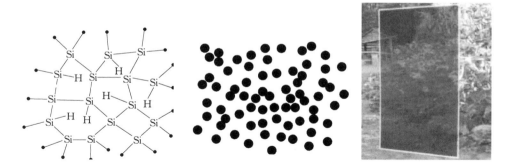

Figure 3.8 Crystal structure of the amorphous crystalline silicon material

less efficient than crystalline ones, with a typical maximum efficiency of about 14%, but they are easier to produce and therefore cheaper. Their low cost makes them ideal for many applications where efficiency is not required and low cost is important. One feature of amorphous silicon solar cells is that their power output diminishes over time, especially for the first few months, after which they are basically stable. The quoted output of the amorphous panel should be produced after this stabilization period.

In conventional monocrystalline silicon and polycrystalline silicon solar cells, p-n junction structure is usually used. However, if a very thin material is used, the light absorption rate will be very low, and the corresponding photo-generated current will also be very small. In order to solve this problem, the amorphous silicon solar cell adopts p-i-n structure or n-i-p structure. The working principle of the amorphous silicon solar cell with the p-i-n structure is that when sunlight hits its surface, it is easy to completely penetrate the p-layer film, so most of the photon absorption occurs in the relatively thick i-layer film. When photons are absorbed by the i-layer film, electron-hole pairs are generated. Then, driven by the electric field between the n-layer and the p-layer, the electrons and holes diffuse to the n-layer and the p-layer, respectively, thus generating photocurrent. This is the most basic power generation principle of the amorphous silicon solar cells.

The amorphous silicon solar cell is convenient for mass production and has great application potential with the advantages of low cost, light weight and high conversion efficiency. However, due to the photoelectric efficiency degradation effect caused by its material, its stability is not high, which directly affects its practical application. In order to improve the performance of the amorphous silicon solar cells, especially improve the conversion efficiency and stability, a-SiGe:H materials with a narrow band gap can be prepared by alloying, such as adding Ge, or by adjusting its internal crystalline phase content to develop nanocrystalline or microcrystalline silicon materials, and using these materials combination to develop multi-junction tandem silicon thin film solar cells. The two most common forms are amorphous silicon/microcrystalline silicon double tandem cell structure and amorphous silicon/amorphous silicon germanium/microcrystalline silicon triple tandem cell structure.

> **Specialized Vocabulary**
> - homogenous 同质的
> - homogenous layer 均质层，均匀层
> - power output 输出功率
> - long-range order 长程有序
> - dangling bonds 悬挂键，悬空键
> - threshold 阈值
> - Plasma Enhanced Chemical Vapor Deposition (PECVD) 等离子体增强化学气相沉积
> - Atmospheric Pressure Chemical Vapor Deposition (APCVD) 常压化学气相沉积
> - evaporation sputter 蒸发溅射

3.1.4 Silicon technology market

Silicon technology currently accounts for about 90% of the world PV market and is a mature technology. Crystalline silicon (c-Si), Czochralski-material, and cast multicrystalline materials are the major contributors to the market, especially crystalline silicon. While the basic design of crystalline devices has remained essentially unchanged over the past two decades, the tremendous growth of the PV market over the previous five years is directly related to the major and dominant position of more efficient, reliable, and lower-cost MC-Si modules.

The advantages of silicon technology are the good availability of the basic material (quartz) for silicon production, sufficient physical properties, which can be used to prepare photovoltaic devices, high energy conversion efficiency and good cell stability during packaging, leading to an expected life of up to 30 years at present. Silicon technology challenges are related to the possible shortage of low-cost solar-grade silicon feedstock, the expense of crystal growth and wafer slicing, and the use of relatively thick (>200 micron) substrates to avoid breakage.

At present, the price of polysilicon has returned to ten years ago, and thin-film cells still maintain the second share as the largest competitor. Thin-film cannot replace crystalline silicon, nor can crystalline silicon replace thin-film. However, when the goal of photovoltaic parity and dual carbon becomes a must, can crystalline silicon still maintain its lead?

> **Specialized Vocabulary**
> - depletion region 耗尽区/层，势垒区，阻挡层
> - ribbon silicon 带硅
> - Czochralski silicon (Cz - Si) 直拉单晶硅
> - cast multicrystalline silicon 铸造多晶硅
> - solar - grade silicon 太阳级硅
> - wafer slicing 晶圆切片，硅片切割
> - feedstock 原料
> - National Renewable Energy Laboratory (NREL) 国家可再生能源实验室

3.2 Inorganic compound solar cells

Compound semiconductor solar cells are composed of two or more semiconductor elements. At present, there are mainly Ⅲ-Ⅴ compound (GaAs) solar cells, Ⅱ-Ⅵ compound (CdS/CdTe) solar cells, ternary (Ⅰ-Ⅲ-Ⅵ) compound (CuInSe$_2$: CIS) solar cells and so on.

3.2.1 Ⅲ-Ⅴ compound solar cells (GaAs)

Gallium arsenide (GaAs) is a typical Ⅲ-Ⅴ compound semiconductor and has the similar Sphalerite crystal structure as silicon. GaAs has a direct band gap, and the band gap width is 1.42 eV (300K). GaAs has high light emitting efficiency and optical absorption coefficient, and it became the foundation material in the optoelectronic field, playing a critical role in the solar cells field.

GaAs single - junction and multi - junction solar cells have the advantages of good spectral responsiveness, long service life in space applications, and high reliability. Despite the high cost, they have been widely used in the space field. Since GaAs solar cells have good radiation resistance and temperature characteristics, they are also suitable for concentrating power generation.

GaAs is often used as a substrate for epitaxial growth of other Ⅲ-Ⅴ semiconductors, including InGaAs and GaInNAs. The photoelectric conversion efficiency of GaAs single - junction solar cells and their tandem cells is the highest among all types of solar cells. However, the GaAs material is expensive, 10 times the price of Si material, and has weak mechanical strength and is fragile, so it is rarely used in the ground field.

A very important application of GaAs is for high efficiency solar cells. GaAs is also known as a thin film single - crystalline device and are high - cost high - efficiency solar cells. In 1970, the first GaAs heterostructure solar cells were created by Zhores Alferov's research team. In the early 1980s, the best GaAs solar cells surpassed silicon solar cells in efficiency, and in the 1990s, GaAs solar cells replaced silicon solar cells as the most commonly used cell type for photovoltaic arrays for satellite applications. Later, dual - and triple - junction solar cells based on GaAs with

Chapter 3 Solar cells

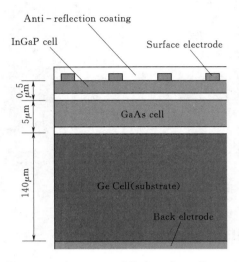

Figure 3.9 Structure of GaAs solar cell

germanium and indium gallium phosphide layers were developed as the basis of a triple-junction solar cell, which held a record efficiency of over 32% and can also operate in the concentrated light of 2,000 suns. Such solar cells power the rovers, Spirit and Opportunity, that explore the Martian surface. In addition, many solar cars use GaAs in their solar arrays. The GaAs-based device holds the world record for the most efficient single-junction solar cell at 28.8%. This high efficiency is attributed to the extremely high-quality GaAs epitaxial growth, surface passivation by the AlGaAs, and the promotion of photon recycling by the thin film design. Also, the complex design of $Al_xGa_{1-x}As$-GaAs devices is very sensitive to infrared radiation. The structure of GaAs solar cell is shown in Figure 3.9.

Specialized Vocabulary

- epitaxial growth   外延生长
- heterostructure   异质结构，异质结
- light emitting efficiency   发光效率，出光效率
- foundation materials   基础材料
- sphalerite crystal structure   闪锌矿晶体结构
- satellite applications   卫星应用
- dual-junction   双结
- triple-junction   三结
- germanium   锗
- indium gallium phosphide   铟镓磷
- rover   火星车
- infrared radiation   红外线照射

3.2.2 Ⅱ-Ⅵ compound solar cells (CdTe/CdS)

Thin film cadmium telluride (CdTe) solar cells are the basis for an important technology that has had a significant commercial impact on solar energy production. Large area monolithic film modules demonstrate long-term stability, competitive performance, and the ability to attract production-scale capital investments.

Over the previous 20 years, significant progress has been made in refining the basic

3.2　Inorganic compound solar cells

CdTe cell structure. The highest current densities achieved were similar to crystalline GaAs when adjusted for small differences in band gap. The open-circuit voltage and fill factor are limited by high forward current recombination and low carrier density, but have nevertheless achieved values about 80% as large as GaAs, again adjusting the bandgap. There are some concerns about the diffusion of copper atoms, but it is unlikely for the well-fabricated cells to achieve significant degradation under normal operating conditions. Although the basic cell-level research should certainly continue, the status of CdTe solar cells is clearly healthy enough to proceed with the mainstream commercialization.

Like GaAs materials, CdTe is very close to the ideal band gap of the photovoltaic materials, and its spectral response is almost the same as the solar spectrum. However, as the temperature changes, the bandgap changes. The CdTe material has a high light absorption coefficient, so it can absorb more than 90% of the sunlight with a thin film of 1μm. CdTe cells usually adopt a heterojunction structure, such as the commonly used CdTe/CdS heterojunction solar cell structure. Therefore, CdTe materials are usually prepared by direct deposition on CdS and other thin film materials. In the preparation process, element diffusion and interface properties are also issues that have to be considered, otherwise it is easy to cause the degradation of solar cell performance.

CdTe cells usually adopt a heterojunction structure, such as the commonly used CdTe/CdS heterojunction solar cell structure. Therefore, when preparing CdTe materials, it is often prepared by the direct deposition on thin film materials such as CdS. In the preparation process, the diffusion of elements and the properties of the interface are also issues that have to be considered, otherwise, the performance of the solar cell may be easily degraded.

CdS is also an important solar cell material. CdS is a direct bandgap photoelectric material with a band width of about 2.4 eV and a high absorption coefficient. It is mainly used as an n-type window material for thin film solar cells. It can form the heterojunction solar cells with thin film materials such as CdTe and $CuInSe_2$ with good performance. CdS materials are rarely used alone as solar cell materials.

The future of CdTe thin-film PV devices in energy production and optical sensing is assured by material properties, lab-scale device performance, and PV module implementation. Enhancing the viability of CdTe/CdS for terrestrial power generation depends on improving the device performance, the cost-effective translation of fabrication processes to the module scale, as well as the cell and module stability, that have been achieved by First Solar. In addition to the efficiency gains in single-junction devices, bandgap tailoring of the absorber by alloying with other group IIB metals can facilitate the development of multi-junction cells with efficiencies approaching 30%. It can significantly reduce the production costs to translate the single-junction efficiency gains from batch processes to continuous module fabrication and to develop monolithic multi-junction modules on a single superstrate. Achieving this goal requires a greater understanding of the relationship

Chapter 3 Solar cells

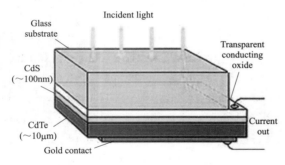

Figure 3.10 Structure of CdTe solar cell

between the processing conditions and the critical material properties needed for high efficiency and good long-term stability. Structure of the CdTe solar cell is shown in Figure 3.10.

In addition to being the basis for single-junction devices, CdTe can also be combined with other IIB-VIA compounds to change their bandgap, allowing the design of multi-junction cells. Multi-junction cell structures using CdTe-based wide bandgap cells in the monolithic structure must deal with the relationship between cell geometry and processing temperature and chemical stability. Materials based on alloys between CdTe and other IIB-VIA compounds allow a wide range of optoelectronic properties to be integrated into devices by design. Semiconductor compounds in the form IIB-VIA provide the basis for the development of tunable materials, which can be obtained by alloying different compounds in pseudo-binary configurations. For photovoltaic heterojunction devices, semiconductors using Cd, Zn, Hg cations and S, Se, Te anions exhibit a wide range of optical bandgaps, indicating their potential for use in optimized devices designed by tailoring material properties. High optical absorption coefficients, 105 cm^{-1}, and direct optical bandgaps of many II-VI semiconductors make them suitable for thin film PV devices. For ground-based PV applications, which require a bandgap of 1.5 eV, considerable progress has been made in the development of solar cells based on CdS-CdTe heterojunctions, where $CdS_{1-y}Te_y$ and $CdTe_{1-x}S_x$ alloys have been shown to play a role in device operation. To develop the next generation of multi-junction cells, a top cell with an absorption bandgap of 1.7 eV is required.

Specialized Vocabulary

- cadmium telluride (CdTe) 碲化镉
- single-junction 单结
- multi-junction 多结
- hetero-junction 异质结
- current density 电流密度
- open-circuit voltage 开路电压
- fill factor 填充因子
- forward-current 正向电流
- carrier density 载流子密度
- soda lime glass 钠钙玻璃
- buffer layer 缓冲层

- close‑space sublimation (CSS) 近空间升华法
- indium tin oxide (ITO) 铟锡氧化物
- optical sensing 光学传感
- terrestrial power generation 地面发电
- cost‑effective 划算的，成本效益好的，合算的
- fabrication processes 制备工艺、合成过程
- single‑junction devices 单结器件
- band gap tailoring 带隙调整
- multi‑junction cells 多结电池
- processing conditions 加工条件，制造条件
- optoelectronic properties 光电性能
- heterojunction devices 异质结器件
- optical absorption coefficients 光吸收系数
- device operation 器件运行，器件工作，设备操作

3.2.3 Ternary compound CuInSe$_2$ solar cells (CIS)

Cu(InGa)Se$_2$‑based solar cells are often touted as one of the most promising solar cell technologies for cost‑effective power generation. This is partly due to the advantages of thin films for low‑cost, high‑rate semiconductor deposition over large areas using layers only a few microns thick and used to fabricate monolithic interconnect modules. Perhaps more importantly, the high efficiency of Cu(InGa)Se$_2$ at the cell and module level has been demonstrated. According to the statistical efficiency of the National Renewable Energy Laboratory (NREL), the current maximum efficiency of the cell is 23.4% with a total area of 0.5 cm^2. Moreover, several companies have demonstrated large‑area modules with efficiencies of more than 12%, including Showa Shell, which confirmed an efficiency of 13.4% for a 3459 cm^2 module. Finally, if fully encapsulated, Cu(InGa)Se$_2$ solar cells and modules have shown excellent long‑term stability in outdoor tests. In addition to potential advantages in large area ground applications, Cu(InGa)Se$_2$ solar cells also show high radiation resistance compared to crystalline silicon solar cells, and can be made into very lightweight with flexible substrates, so they are also promising for space applications.

In a typical CIGS solar cell structure, in addition to glass or other flexible substrate materials, there are seven thin film materials such as bottom electrode Mo layer, CIGS absorption layer, CdS buffer layer (or other cadmium free materials), I‑ZnO and Al‑ZnO window layers, MgF$_2$ antireflective layer and top electrode Ni/Al. As shown in Figure 3.11.

Since its earliest development, CuInSe$_2$ has been considered a promising solar cell because of its favorable electronic and optical properties, including a direct bandgap with high absorption coefficients and an inherent p‑type conductivity. With the development of science and technology, it has become obvious that it is a very tolerant material, because:

Chapter 3 Solar cells

Figure 3.11 Structure of Cu(InGa)Se$_2$ solar cell

① High-efficiency devices can be made with a wide tolerance to changes in Cu(InGa)Se$_2$ composition; ② Grain boundaries are inherently passive, so even films with grain size less than 1 μm can be used; ③ The device behavior is not sensitive to lattice mismatch between Cu(InGa)Se$_2$ and CdS or junction defects caused by impurities. The structure of the Cu(InGa)Se$_2$ solar cell is shown in Figure 3.11. However, there has always been concern about susceptibility to grain boundary degradation by moisture.

The understanding of Cu(InGa)Se$_2$ films used in PV devices is mainly based on the study of its base material, pure CuInSe$_2$. However, the material used to make solar cells is Cu(InGa)Se$_2$, which contains a large amount (on the order of 0.1%) of sodium. Although the behavior of CuInSe$_2$ provides a good basis for understanding device quality materials, there are clear differences in the behavior of CuInSe$_2$ when Ga and Na are present in the film. Recently, Cu(InGa)Se$_2$ has been reviewed in the context of solar cells with a focus on electronic properties.

Deposition of CIGS thin films can be achieved by evaporation, sputtering and CVD-based techniques, which are carried out under high vacuum. There are also printing method using nanoparticles (inks) and liquid phase deposition method under atmosphere condition at 600~700 ℃.

CIGS thin-film solar cells have strong radiation resistance, good cell stability, basically no attenuation, good low-light characteristics, and strong competitiveness as a space power source, so it is expected to become one of the mainstream products of the new generation of solar cells. However, because the CIS (CIGS) thin film material is composed of multiple elements, the element ratio is sensitive, the multiple crystal structure is complex, and it is difficult to match with the multi-layer interface, which requires high precision, repeatability and stability of material preparation. Therefore, the technical difficulty of material preparation is relatively high.

3.3 Novel organic semiconductor thin film solar cells

> **Specialized Vocabulary**
> - high-rate semiconductor deposition 高效/高速半导体沉积
> - radiation resistance 耐辐射性, 辐射抗性
> - absorption coefficient 吸收系数
> - lattice mismatch 晶格失配
> - aluminium doped zinc oxide (AZO) 掺铝氧化锌
> - indium tin oxide (ITO) 铟锡氧化物
> - fluorine doped tin oxide (FTO) 掺氟氧化锡
> - chemical bath deposition 化学浴沉积法
> - sputtering 溅射
> - selenization 硒化法
> - intermetallic 金属间化合物
> - liquid phase deposition 液相沉积法
> - radio Frequency Sputtering 射频溅射

In summary, the general compound semiconductor materials are direct bandgap materials with high light absorption coefficient. Therefore, only a few microns thick material is required to fabricate high-efficiency solar cells. In order to fully absorb the energy of sunlight, people tend to choose compound semiconductor materials with different bandgaps to form efficient tandem solar cells, which is expected to gradually improve the photoelectric conversion efficiency and increase its stability.

3.3 Novel organic semiconductor thin film solar cells

3.3.1 Perovskite solar cells

In 2009, Miyasaka research group introduced perovskite materials ($CH_3NH_3PbI_3$ & $CH_3NH_3PbBr_3$) as light absorption film into the field of solar cells for the first time with the photoelectric conversion of 3.8%. Since then, organometal halide perovskites have emerged as a promising light-harvesting material for high-efficiency solar cell devices. Perovskite sensitizer made a breakthrough in solid-state solar cells, where a rapid increase in efficiency has achieved at least 25.7%, while the highest laboratory conversion efficiency of the perovskite-silicon tandem cells has reached 29.5%. Perovskite solar cells possess two important characteristics of low cost and high efficiency, which provides a new way to utilize solar energy. As the most talked about, the most promising solar cell, its in-depth study also has important strategic significance.

An inorganic-organic lead halide perovskite is any material that crystallizes into an ABX_3 [where A is an organic ammonium cation ($CH_3NH_3^-$), B is Pb or Sn, and X is a halide anion such as Cl^-, Br^-, or I^-] structure (Figure 3.12). The size of the cation A is critical for the formation of the tightly packed perovskite structure. In particular, the cation A must fit into the space composed of four adjacent corner-sharing BX_6 octahedra. Among various inorganic-organic lead halide perovskite materials, methylammonium lead iodide ($MAPbI_3$) has a band gap of about 1.5~1.6 eV and an optical absorption spectrum up to 800 nm, which is widely used as a light harvester in perovskite solar cells.

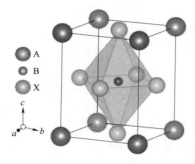

Figure 3.12　Crystal structure of the perovskite absorber adopting the perovskite ABX_3 form, where A is methylammonium, B is Pb and X is halogen[1]

Generally, perovskite solar cells are composed of a transparent conductive substrate, a dense barrier layer, a porous electron transport layer, a perovskite absorption layer, a hole transport layer and a metal electrode, as shown in Figure 3.13. The material composition, microstructure and properties of the electron transport layer, the perovskite absorption layer and the hole transport layer have a significant effect on the photovoltaic performance and long term stability of the perovskite solar cells. At present, the perovskite solar cells have three main alternative structures. The first is mesoporous perovskite solar cells which commonly use the mesoporous TiO_2 material as the electron transport layer. By the impact of photons, there will be exciton separation phenomenon to form electrons and holes in the cell. Affected by the light-absorbing material, TiO_2 will transfer the electrons to the FTO surface. The second structure is planar perovskite solar cells. This type of perovskite solar cells is mainly developed for low temperature manufacture, with the advantages of low cost. When exposed to light, the cell will produce electron-hole pairs, thus making the electrons more active. When the circuit is connected, the electron-hole pairs will move rapidly, resulting in the current, and then give full play to the role of perovskite solar cells. The third is flexible perovskite solar cells. Until 2015, the research of flexible perovskite solar cells in the world mainly used the atomic layer vapor deposition manufacturing method, which has boosted the efficiency of flexible perovskite solar cells to 18%. The third type is the perovskite solar cell with inverse structure,

---

[1] Mingzhen Liu, Michael B. Johnston, Henry J. Snaith. Efficient planar heterojunction perovskite solar cells by vapour deposition [J]. Nature, 2013, 501 (7467): 395-8.

3.3 Novel organic semiconductor thin film solar cells

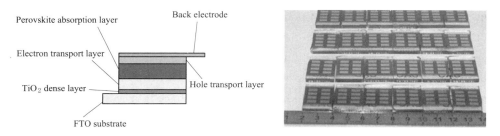

Figure 3.13 Schematic diagram of the perovskite solar cell structure and the appearance of the cells

which has the opposite structure to that of the planar perovskite solar cell.

The working principle of the perovskite solar cell is shown in Figure 3.14. Methyl ammonium (MA) lead halide perovskite absorbs photons under light, and the electrons in the valence band jump to the conduction band; then the conduction band electrons are injected into the conduction band of TiO_2 and then transfer to the FTO conductive substrate. Meanwhile, holes are transferred to the organic hole transport layer, thus the electron–hole pairs are separated. When the external circuit is connected, the movement of electrons and holes will generate current. The main function of the dense layer is to collect the electrons which are injected into the electron transfer layer, resulting in the separation of the electron–hole pairs from the absorption layer. The main function of the absorption layer is to absorb the electron–hole pairs generated by sunlight and transfer them efficiently; thus, electrons and holes reach respectively the corresponding dense layer and the hole transport layer. The main role of the hole transport layer is to collect and transfer holes which are injected from the absorption layer, and in the meantime to accelerate separation of electron–hole pairs in the absorption layer together with the dense layer.

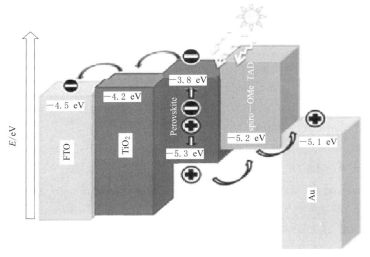

Figure 3.14 Working principle of perovskite solar cell

Chapter 3 Solar cells

On the preparation methods of perovskite solar cells, there are one-step spin-coating method, two-step sequential method, and vapor deposition in a high-vacuum chamber, plus other variants. Additionally, the fabrication methods for different device architectures such as planar and mesostructured cells are a little bit different.

At present, the development of the perovskite solar cells is very vibrant, but there are still a number of key factors that may restrict the prospects of perovskite solar cells. First is the stability of the solar cell. The perovskite solar cells are sensitive and unstable in the air environment with sensitivity to oxygen and moisture, and the attenuation of efficiency for the MA compositions is rapid and significant. Second, the perovskite absorption layer contains soluble heavy metal lead, which is easy to cause environment pollution, though the concentration is low. Third, the most widely used fabrication method of perovskite materials is spin-coating, which is difficult to deposit in large areas and as a continuous film, so other methods need to be improved in order to obtain the large-area high-efficiency perovskite solar cells to enable the future commercial production. Finally, the theoretical study of the perovskite solar cells has yet to be strengthened.

> **Specialized Vocabulary**
> - organometal halide perovskites   有机金属卤化物钙钛矿
> - inorganic-organic lead halide perovskite   有机—无机卤化铅钙钛矿
> - dense layer   致密层
> - electron transport layer   电子传输层
> - perovskite absorption layer   钙钛矿吸收层
> - hole transport layer   空穴传输层
> - metal electrode   金属电极
> - back electrode   背电极
> - mesoporous perovskite solar cells   介孔钙钛矿太阳电池
> - planar perovskite solar cells   平面/板钙钛矿太阳电池
> - flexible perovskite solar cells   柔性钙钛矿太阳电池
> - spin-coating   旋涂,旋涂仪
> - atomic layer vapor deposition   原子层气相沉积

3.3.2 Organic polymer solar cells

The research on organic solar cells began in 1959, initially with a simple sandwich structure, and now it has developed into the bulk heterojunction-blended polymer solar cells. The structure of an organic solar cell has a great influence on its performance, such as the absorption of light, the generation and separation of the excitons in the cell, as well as the electron and hole transport. Therefore, designing a rational organic solar cell structure can improve the performance of the cell. The bulk heterojunction structure greatly increases the contact interface for the acceptor, effectively improves the separation

3.3 Novel organic semiconductor thin film solar cells

efficiency of the excitons, and further improves the photoelectric conversion efficiency, thus opening up the research direction of polymer solar cells. At present, the laboratory photoelectric conversion efficiency of the organic solar cells has reached 18.2%.

Organic solar cells (Figure 3.15) can be colored, transparent, and applied to flexible, lightweight films. They can generate electricity even under cloudy skies. It's a kind of polymer solar cell that uses organic electronics, a branch of electronics that deals with conductive organic polymers, or small organic molecules, used for light absorption and charge transport to generate electricity from sunlight through PV effect.

The structure of the polymer solar cell mainly includes single layer polymer solar cells, double layer polymer solar cells (p - n heterojunction structure) and bulk heterojunction polymer solar cells. Later, inverted

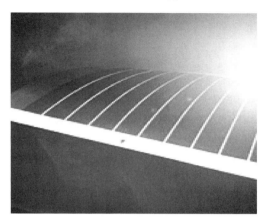

Figure 3.15 Organic solar cell

polymer solar cells and tandem polymer solar cells emerged. At present, the bulk heterojunction structure has become the mainstream of the polymer solar cells. It mainly includes ITO anode, PEDOT: PSS anode modification layer, bulk heterojunction active layer, LiF cathode modification layer and aluminum cathode. The electron donor (D) and electron acceptor (A) materials are in close contact and interpenetrate to form an interpenetrating network - like continuous structure, so the biggest advantage of this structure is to increase the contact area of donor/acceptor material. Meanwhile, in the active layer of the bulk heterojunction structure, most excitons can reach the donor/acceptor interface within the diffusion length and charge separation occurs there, thus greatly improving the separation efficiency of the excitons.

The working principle of the bulk heterojunction polymer solar cells: In bulk heterojunction solar cells, the incident photons are absorbed by the donor and acceptor materials of the active layer to generate excitons (electron - hole pairs in bound state). After these excitons diffuse to the interface of the donor - acceptor, driven by the electron energy level difference of the donor - acceptor, the excitons undergo charge separation to generate electrons at the lowest unoccupied molecular orbital (LUMO) energy level of the acceptor, and holes at the highest occupied molecular orbital (HOMO) energy level of the donor. The separated electrons and holes are transported to the cathode and the anode respectively along the acceptor and the donor in the active layer under the action of the built - in electric field of the device, and are collected by the two electrodes to form the photocurrent and photovoltage. This is how bulk heterojunction polymer solar cells work. The bulk heterojunction structure can solve the problem of short exciton diffusion distance in organic semiconductor materials, and can obtain higher photocurrent.

There are different types of junctions for organic PV cells. Among them, single layer organic photovoltaic cells are the simplest of the various forms of organic photovoltaic cells. These cells are made by sandwiching a layer of organic electronic materials between two metallic conductors, typically a layer of indium tin oxide (ITO) with high work function and a layer of low work function metal such as Al, Mg or Ca. The basic structure of such a cell is illustrated in Figure 3.16 (a). In practice, such single-layer organic PV cells have low quantum efficiency ($<1\%$), low power conversion efficiency ($<0.1\%$) and poor working effect. A major problem with them is that the electric field resulting from the difference between the two conductive electrodes is seldom sufficient to break up the photo-generated excitons. Usually the electrons bind to the hole instead of reaching the electrode. To solve this problem, multi-layer organic PV cells have been developed. This organic PV cell contains two distinct layers between conducting electrodes [Figure 3.16 (b)]. There are differences in electron affinity and ionization energy between the two layers. As a result, electrostatic forces are generated at the interface between the two layers. The materials are chosen to make the differences large enough, so these local electric fields are strong, which may break up the excitons much more efficiently than the single layer photovoltaic cells do. The layer with higher electron affinity and ionization potential is the electron acceptor, and the other layer is the electron donor. This structure is also known as a planar donor-acceptor heterojunction. The diffusion length of excitons in organic electronic materials is usually on the order of 10 nm. The thickness of the layer should also be in the same range as the diffusion length in order for most excitons to diffuse to the interface of the layer and break up into charge carriers. However, polymer layers typically need to be at least 100 nm thick to absorb enough light. At such a large

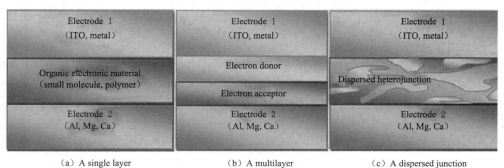

Figure 3.16 Sketch of a single layer, a multilayer and a dispersed junction photovoltaic cell[1]

---

[1] Jenny Nelson. Organic photovoltaic films [J]. Current opinion in solid state and materials science, 2002, 6: 87-95.

thickness, only a small fraction of excitons can reach the heterojunction interface.

In order to solve this problem, a new type of heterojunction photovoltaic cell, namely dispersed heterojunction PV cell, is designed, which is the dispersed heterojunction photovoltaic cells. In this type of PV cell, the electron donor and acceptor are mixed together to form a polymer blend [Figure 3.16 (c)]. If the length scale of the blend is similar to the exciton diffusion length, most of the excitons produced in the two materials are likely to reach the interface, where the excitons break efficiently. Electrons move to the acceptor domain and are carried through the device then collected by an electrode, and holes are pulled in the opposite direction and collected on the other side.

Most plastics used in organic solar cells are cheap to produce and can be produced in large quantities. Combined with the flexibility of organic molecules, organic solar cells are potentially cost effective in certain PV applications. Molecular engineering, such as changing the length and functional groups of polymers, can alter the energy gap, resulting in chemical changes in these materials. Organic molecules have a high optical absorption coefficient, thus a small amount of material can absorb a large amount of light. Compared with rigid inorganic PV cells, the main disadvantages of organic PV cells are low efficiency, low stability and low strength.

The difficulties associated with organic PV cells include the low external quantum efficiency (up to 70%) compared to inorganic PV devices, which is mainly due to the large bandgap of organic materials. Antioxidant and reduction instability and recrystallization are problems. Temperature changes can also cause device degradation and decreased performance over time. This occurs to different extents for devices with different compositions, and this is an area into which active research is taking place. Other important factors include exciton diffusion length, charge separation and collection, as well as charge transport and mobility, which are affected by the presence of impurities.

With the continuous deepening of the organic solar cell research, organic solar cells with new structural characteristics will continue to appear, and new application value will also continue to emerge.

> **Specialized Vocabulary**
> - direct current 直流，直流电
> - exciton 激子
> - heterojunction 异质结
> - conjugated system 共轭体系
> - highest occupied molecular orbital (HOMO) 最高占有分子轨道
> - lowest unoccupied molecular orbital (LUMO) 最低空分子轨道
> - covalent bond 共价键

- hydrocarbons 碳氢化合物，烃类
- delocalized bonding 离域键，非定域键
- antibonding orbital 反键轨道
- planar donor - acceptor heterojunction 平面供体—受体异质结
- diffusion length 扩散长度

3.3.3 Quantum dot solar cells (QDSCs)

At present, there are quantum dot sensitized solar cells, quantum dot polymer hybrid solar cells, quantum dots schottky and depletion heterojunction quantum dot solar cells, etc. The highest efficiency of quantum dot sensitized solar cells is currently up to 13.4%.

The structure and working principle of quantum dot sensitized solar cells are similar to those of dye - sensitized solar cells. As shown in Figure 3.17, it mainly consists of a conductive substrate material (transparent conductive glass), wide bandgap oxide semiconductor nanoporous film, quantum dots, electrolyte and counter electrode. Under the sun light, the quantum dots absorb the photons with the energy greater than the bandgap width, which causes them to jump from the ground state to the excited state. The electrons in the excited state will be quickly injected into the conduction band of the nano wide bandgap semiconductor, and then diffuse towards and transport to the conductive substrate, arriving at the counter electrode through the external circuit. The quantum dots in the oxidized state are reduced and regenerated by the reducing agent in the electrolyte, and the oxidant in the electrolyte is reduced by receiving electrons at the counter electrode. The above is a cycle of electrons during the work mode of solar cells. However, in

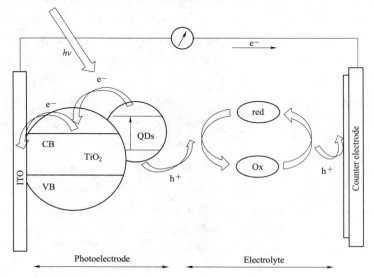

Figure 3.17 Structure and working principle of quantum dot solar cells

addition to this ideal electronic transmission, there is an inverse electron recombination process when the solar cells are actually working, which is a limit to the improvement of the cell performance.

> **Specialized Vocabulary**
> - quantum dot sensitized solar cells 量子点敏化太阳电池
> - quantum dot polymer hybrid solar cells 量子点聚合物杂化太阳电池
> - quantum dot schottky solar cells 量子点肖特基太阳电池
> - depletion heterojunction quantum dot solar cells 耗尽异质结量子点太阳电池
> - inverse electron recombination 电子反向复合过程

3.3.4 Dye-sensitized solar cells (DSSCs)

The modern version of dye-based solar cells, also known as the Grätzel cell, was originally co-invented by Brian O'Regan and Michael Grätzel at the University of California, Berkeley, in 1988. Since 1991, when Grätzel and coworkers reported the development of the highly efficient, novel DSSCs, researchers throughout the world have intensively studied the mechanisms, new materials and commercialization of DSSCs. In the laboratory, the maximum efficiency under AM1.5 achieved 13%. Furthermore, the sealed cells has achieved satisfactory long-term stability under relatively mild test conditions (low temperature and no UV exposure). Commercial DSSC production can be achieved for indoor applications such as calculators and several types of watches. In addition, recently, a new DSSCs design achieved a power conversion efficiency of up to 28.9% for indoor lighting under 1,000 lux.

1. *What is DSSC?*

DSSC is an attractive and promising solar device that has been extensively studied worldwide and its photovoltaic mechanism is widely understood. It is based on a semiconductor formed between a photosensitized anode and an electrolyte, a photochemical system. Commercial applications of DSSC have been intensively studied for many years. The cost of commercially manufacturing DSSC is expected to be relatively low because the cells are made of low-cost materials and assembly is simple, and do not require high vacuums or high temperatures.

When it falls onto the DSSC, the light is absorbed by the dye. The electrons that are excited, due to the extra energy the light provides, can escape from the dye and into the TiO_2 and diffuse through the TiO_2 to the electrode. They are eventually returned to the dye through the electrolyte. Therefore, DSSC uses a dye molecule to absorb light, similar with chlorophyll in photosynthesis. But the "chlorophyll" is not involved in charge transport. It just absorbs light and generates a charge, and then those charges are conducted by some well-established mechanisms. That is exactly what DSSC system does.

2. The structure of DSSC

The structure of DSSC mainly includes three parts, porous anode film on the conductive substrate which adsorbed dyes, electrolyte and counter electrode as a sandwich structure (Figure 3.18). The most important part is the porous photo-anode film where the electronic conduction takes place.

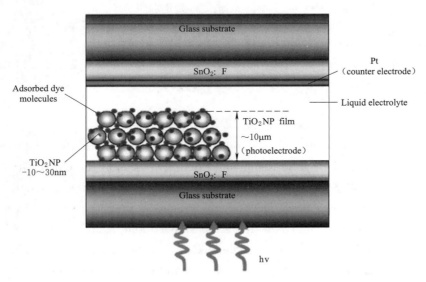

Figure 3.18 Structure of DSSC

3. The primary processes and mechanism of DSSC

The schematic working diagram of DSSC is shown in Figure 3.19. The following primary steps convert photons to current. First, the photosensitizers adsorbed on the TiO$_2$ surface absorb incident photons, then the photosensitizers are excited from the ground state (S) to the excited state (S*) owing to the MLCT (Metal to Ligand Charge Transfer) transition. Second, the excited electrons are injected into the conduction band of the TiO$_2$ electrode, resulting in the oxidation of the photosensitizer (S$^+$).

$$S + h\nu \rightarrow S^* \quad (3.1)$$
$$S^* \rightarrow S^+ + e^- (TiO_2) \quad (3.2)$$

Injected electrons in the conduction band of TiO$_2$ are transported between TiO$_2$ nanop-

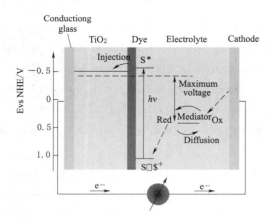

Figure 3.19 Schematic working diagram of DSSC[1]

---

[1] Micheal Gratzel. Dye-sensitized solar cells [J]. Journal of photochemistry and photobiology C: photochemistry reviews, Michael Gratzel 2003, 4: 145-153.

articles with diffusion toward the back contact (TCO) and consequently reach the counter electrode through the external load and wiring. Third, the oxidized photosensitizer (S^+) accepts electrons from the I^- ion redox mediator, regenerating the ground state (S), and I^- is oxidized to the oxidized state, I_3^-.

$$S^+ + e^- \rightarrow S \tag{3.3}$$

Then, the oxidized redox mediator, I_3^-, diffuses toward the counter electrode and is reduced to I^- ions.

$$I_3^- + 2e^- \rightarrow 3I^- \tag{3.4}$$

Overall, electric power is generated without permanent chemical transformation.

In contrast to conventional p – n type solar cells, the mechanism of DSSCs does not involve the charge recombination process between electrons and holes, because electrons are only injected from the photosensitizer into the semiconductor, and no holes are formed in the valence band of the semiconductor. In addition, charge transport occurs in TiO_2 films separated from photon absorption sites (i. e. photosensitizers), thus, efficient charge separation is expected. This photon – to – current conversion mechanism in DSSCs is similar to the photosynthesis mechanism in nature, where chlorophyll acts as a photosensitizer and charge transport occurs in the membrane.

4. *Advantages*

The biggest advantage of DSSC is that the charge transfer is accomplished by majority carrier transmission charge, and there is no traditional p – n junction solar cell in minority carrier and charge transport materials surface recombination. In addition, its preparation process is simple, with simple requirements for the environmental conditions. Moreover, it has a potentially low cost (1/10~1/5 of silicon solar cells), a long service life, such as 15~20 years in lower light applications, and it can be mass – produced because of the simple structure and easy manufacture. In conclusion, the advantages are simple preparation and less demanding on the environment.

5. *Problems*

For expanded commercial applications, however, there are several problems for DSSC. Overcoming these problems would bring DSSC closer to expanded commercialization. First of all, the efficiency of DSSC should be improved. For commercial applications, the full sun efficiency which is 12.3% now needs to be improved. Therefore, expanding the absorption property of the photosensitizer to near IR region is necessary to increase J_{sc} and resulting in efficiency improvement. Second, for outdoor applications, DSSC should also have long – term stability under more rigorous conditions, since it has already been tested only under relatively mild conditions. Third, development of solid – state electrolytes in DSSC is considered essential for developing a cell with long – term stability and is therefore critical for commercialization. However, it is very difficult to form a good solid-solid interface.

> **Specialized Vocabulary**
> - anode/photoanode film 阳极/光阳极薄膜
> - conductive substrate 导电基底
> - counter electrode 对电极
> - mesoporous 介孔，中孔
> - anatase 锐钛矿（型）
> - nanocrystalline 纳米晶（体）
> - photosensitizers 光敏剂
> - electrolyte 电解液，电解质
> - redox 氧化还原，氧化还原剂，氧化还原反应
> - charge recombination 电荷复合
> - valence band 价带

3.4 Concentrator solar cells

Photovoltaic concentrators are based on the principle of the use of a lens or mirror to collect the sun's power onto a small photovoltaic device in order to increase the energy conversion efficiency with the same cell active area and substitute the expense of a larger solar cell by a cheaper optical system (Figure 3.20). This allows reducing the cell area required to produce a certain amount of power. The goal is to significantly reduce the cost of electricity generation by replacing expensive PV converter areas with cheaper optical materials. This approach also offers the opportunity to use higher performance PV cells that would be prohibitively expensive without concentrations. As a result, the concentrator module can easily exceed 30% or even 40% energy conversion efficiency. In the future, the use of multi-junction cells is expected to increase this to more than 50%. While the concept is simple and has been tested since the earliest interest in ground-based PV, the practice has proven to be deceptively difficult.

Concentrator research has focused on the PV cell itself, which is now well developed and commercialized. However, the major remaining technical obstacles are the difficult cell packaging requirements due to high heat flux and current densities, coupled with the need for more economical and reliable tracking systems and module designs. The major market barriers are due to the concentrating systems, which in most cases have to track the sun, not being well suited to the existing PV market, serving small remote loads and, more recently, building integrated applications. Another disadvantage of concentrator system is the limitation to locations with a high proportion of direct radiation, where solar tracking systems

3.4 Concentrator solar cells

are required due to small aperture sizes. Marketing of PV concentrators has not been successful.

In static concentrator designs that do not require solar tracking, the concentration ratio (ratio of the module aperture area to cell area) varies from 2 to 4 and exceeds 1,000 times in some two-axis tracking systems. The means of optical concentration includes various two-axis and one-axis reflection and refraction methods, as well as many new means, such as luminescence and holographic concentrators. The development of concentrators has been aided and influenced by the parallel development of materials and other technologies. Finding and tracking the sun, for example, once a hassle, is now relatively simple, thanks to very low-cost computing technology and global positioning systems.

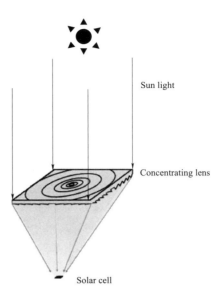

Figure 3.20 Concentrator solar cells

On the other hand, developments in the global semiconductor industry are often directly applied to concentrated cells. Examples include larger wafers, improved processing equipment, the emergence of organo-metallic chemical vapor deposition (OMCVD) for the fabrication of multi-junction III-V cells, and improved packaging materials with superior thermal properties. Many technical problems facing the further development of concentrators can be regarded as material problems. These include the development of polymer reflectors with improved weather resistance and low-cost molding methods for Fresnel lenses. In other words, the development of the concentrator takes place in the larger technical domain. The development of new materials and technologies can come from any direction and make previous dreams possible. Unfortunately, tracking the sun is still influenced by obvious the 19th Century gear and motor technology. The need to track the sun remains the Achilles' heel of PV concentrators.

A key issue for PV concentrators is the need to develop energy efficient and cost-effective solar cells that can operate at high solar fluxes (>500 suns) and provide economical concentrator systems. In addition, the PV concentrator market is yet to be developed. Concentrators are considered tools for generating large amounts of non-polluting renewable energy. As of now, costs are still too high to compete with fossil fuel generation, or even with the most direct renewable competitors, wind and solar. However, the cost gap is narrowing and it seems likely that concentrator systems will find cost-effective applications in the future.

In summary, the types of solar cells emerge in an endless stream, each with its own characteristics. It is foreseeable that the efficiency of solar cells will become higher and

Chapter 3 Solar cells

higher, that the cost will gradually decrease, that the structure will continue to mature, and that the technical process will continue to improve. If the function and structure of each type of solar cell can be properly utilized, the utility of solar photoelectric conversion can be exerted more widely.

Specialized Vocabulary

- concentrator solar cell 聚光太阳电池
- concentration ratio 聚光比，聚光比例
- fresnel lenses 菲涅尔透镜
- photovoltaic concentrators 光伏聚光器
- lenses 透镜
- mirrors 反射镜，反光镜
- optical lens 光学透镜
- active area 有效面积
- technical barriers 技术瓶颈，技术壁垒
- heat flux 热通量，热流
- cell packaging 电池封装，电池包装
- sun-tracking system 太阳跟踪系统
- two-axis tracking systems 双轴跟踪系统
- one-axis 单轴
- reflective 反射，反光
- refractive 折射，折射的
- luminescent 发光，发光的，冷光的，场致发光
- holographic concentrators 全息聚光，全息聚光器
- Global Positioning System (GPS) 全球定位系统
- Organo-Metallic Chemical Vapor Deposition (OMCVD) 有机金属化学气相沉积
- polymer reflectors 聚合物反射镜
- molding methods 成型方法
- sunlight flux 阳光通量
- fossil fuel-fired generation 化石燃料发电

Reading Materials

1. *Perovskite-based tandem solar cells*
2. *Silicon solar cell production*

Reading Materials

3.4 Concentrator solar cells

Exercises and Discussion

1. What are the advantages and disadvantages of the monocrystalline, multicrystalline and amorphous crystalline silicon?

2. What do you think of the future development of silicon-based solar cells?

3. Please describe the structure and working mechanism of CdTe, CIGS and GaAs solar cells, respectively.

4. Which is the important component in concentrator solar cells? Why?

5. What is the big difference between DSSC and perovskite solar cells?

Chapter 4
PV Modules

4.1 The definition and types of PV modules

A PV module is a minimum combination of the integral solar cell device that has external packaging and internal connection, and can provide an individual DC output. The PV module is formed by the encapsulation of multiple interconnected single solar cells. It is the core and the most important part of the solar power generation system.

PV modules consist of many interconnected solar cells encapsulated into a single, durable and stable unit. The key purpose of encapsulating a set of electrically connected solar cells is to protect them and their interconnecting wires from the harsh environment in which they are used. For example, solar cells, because they are relatively thin, are vulnerable to mechanical damage if not protected. In addition, the metal grid on the top surface of the solar cells and the wires interconnecting the individual cells will be corroded by water or steam. Two key functions of the encapsulation are to prevent mechanical damage to the solar cell and to prevent water or steam from corroding the electrical contacts. Moreover, in order to meet the requirements of the load, series and/or parallel connection is required to form a minimum unit to be used as power supply independently. The single solar cell needs to supply electric power when in series or parallel connection because the voltage output of the single solar cell is hard to meet the normal demand.

There are several types of PV modules. According to the types of the solar cells, there are crystalline silicon (monocrystalline or multicrystalline silicon) solar cell modules, amorphous silicon thin film solar cell modules and gallium arsenide solar cell modules, to name but a few (Figure 4.1). According to the encapsulation materials and process, there are epoxy resin encapsulation panels and laminated packaging solar cell modules. According to the application, there are standard solar cell modules and building material-type solar cell modules. Among them, building material-type solar cell modules are divided into single glass with back face opaque to transparent, double-sided laminated glass module components and double-hollow glass module components. The modules based on crystalline silicon solar cells account for more than 85% of the market.

4.2 Module structure

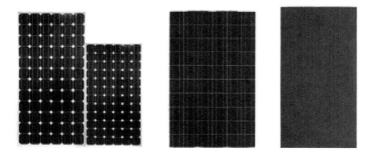

Figure 4.1 Images of monocrystalline silicon modules, polycrystalline silicon modules and amorphous silicon modules

Specialized Vocabulary
- modules 组件，模组，模块
- epoxy resin encapsulation 环氧树脂封装
- laminated packaging 层压封装

4.2 Module structure

Most bulk silicon PV modules consist of a transparent top surface, an encapsulant, a rear layer and a frame around the outer edge. In most modules, the top surface is glass, the encapsulant is ethyl vinyl acetate (EVA) and the rear layer is Tedlar, as shown in Figure 4.2.

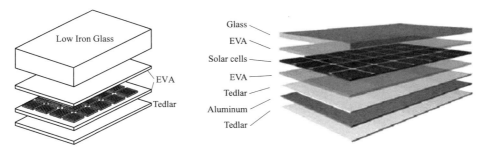

Figure 4.2 Typical bulk silicon module materials

The front surface of the PV module must have high transmittance in the wavelengths that can be used by the solar cells in the module. For silicon solar cells, the top surface must have high transmission of light in the wavelength range from 380 nm to 1200 nm. In addition, reflections from the front side should be low. There are several options for top surface materials, including acrylic, polymer and glass. Tempered glass with low iron content is the most commonly used glass because it is inexpensive, strong, stable, high-

Chapter 4　PV Modules

ly transparent in the visible range, impervious to water and gas, and has good self-cleaning properties. Encapsulants are used to provide adhesion between the solar cells, the top and back surfaces of the photovoltaic modules. This encapsulant should be stable at high temperatures and high UV exposure. It should also be optically transparent and should have a low thermal resistance. EVA is the most commonly used encapsulant material. EVA are thin sheets that are inserted between the solar cell and the top and rear surfaces. This sandwich is then heated to 150 ℃ to polymerize the EVA and bond the modules together. The key feature of the rear surface of a PV module is that it must have low thermal resistance and must prevent the entry of water or steam. In most modules, a thin polymer sheet (usually Tedlar) is used as the rear surface. Some photovoltaic modules, called double-sided modules, are designed to receive light from the front or back of the solar cell. In a bifacial module, both front and back must be optically transparent. The final structural component of a module is the edge or frame of the module. Conventional photovoltaic module frames are usually made of aluminum. The frame structure should be free of projections that could cause the inclusion of water, dust or other material. The PV module consists of a transparent front, encapsulated solar cell and a back. Although the front material (superstrate) is usually low-iron, tempered glass, for some special module types, some other front materials used include polycarbonate or non-toughened glass. For flexible modules, ethylene tetrafluoroethylene (ETFE), a fluorine-based plastic, with high corrosion resistance and strength over a wide temperature range, is often used. The back is usually opaque and the most commonly used material is polyvinyl fluoride (PVF). Transparent back is also possible—transparent back materials are often used in modules to integrate into the building envelope (facade or roof). Between the rear glass and the solar cell, the encapsulation material is placed. Many different materials are available for encapsulation, but the two most commonly used are EVA and polyvinyl butyral (PVB). In the automotive industry, PVB is also used as a laminate for safety windscreens. It is used as an encapsulation material for transparent modules. EVA is used to encapsulate cells in standard modules. Other less common encapsulation materials are thermoplastic polyurethane (TPU) and castable polyurethane or silicone casting resins, for example for transparent modules or other demanding applications. International standards specify required mechanical properties (e.g. impact resistance) and module qualification procedures.

　　Standard PV modules are laminates composed of a 4 mm thick glass superstrate, an 0.5 mm thick encapsulating polymer layer (EVA), a ~ 0.18 mm thick layer of Si solar cells, another layer of EVA with the same thickness as the previous one, and finally a 0.1 mm thick thin multi-layered back sheet made of Tedlar/Aluminum/Tedlar, as shown in Figure 4.2. The majority of solar cells available on the market are made of either mono or polycrystalline Si between thin layers of EVA in the plane of the cells. Two main conductors, called busbars, connect the cells together and are placed on the upper and the lower sides of the cells.

4.3 Encapsulation technology

It is difficult to meet the demand of conventional electricity due to the very small output power of single solar cell. Therefore, the solar cells need to be encapsulated into modules to improve the power output. Encapsulation is a critical step in solar cell manufacturing. The encapsulation of the cells is not only the process for the module but also the guarantee of cell lifetime, enhancing the resistance of the cell to degradation.

EVA encapsulation technology is the most popular. The encapsulation process is: first, use glass as the substrate, and then place EVA films on both sides of the solar cell. The encapsulation process involves heating to a certain temperature under vacuum, so that the EVA softens, and as the temperature decreases, solidification occurs, binding the solar cells in a fixed pattern. The second stage involves coating the connector materials at the edge of the base plate and the top plate of the module, and adding the frame. The third phase involves release testing of the modules, cleaning, packing, and putting in storage. The schematic diagram of the encapsulation process is in Figure 4.3.

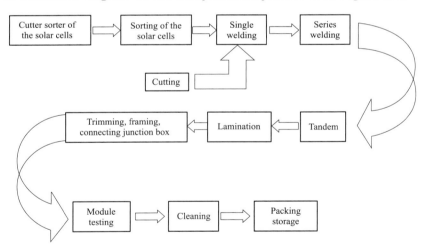

Figure 4.3　EVA copolymer encapsulation process

Specialized Vocabulary

- encapsulant　密封材料，密封剂
- a rear layer　底板
- ethyl vinyl acetate (EVA)　醋酸乙烯乙酯
- Tedlar　泰德拉（宇宙飞船用以保留热量的绝缘材料），聚氟乙烯

Chapter 4 PV Modules

- ethylene tetraflaoroethylene (ETFE)　乙烯四氟乙烯
- polyvinyl fluoride (PVF)　聚氟乙烯
- encapsulation material　封装材料
- busbars　母线
- superstrate　覆盖物
- polyvinyl-butiral (PVB)　聚乙烯醇缩丁醛
- thermoplastic polyurethane (TPU)　热塑性聚氨酯
- power output　输出功率

Reading Materials

1. *Review on feasible recycling pathways and technologies of solar photovoltaic modules*
2. *Solar photovoltaic module production: Environmental footprint, management horizons and investor goodwill*

Reading Materials

Exercises and Discussion

1. Why do we have to encapsulate the solar cells to produce a module?
2. What is the structure of the module?
3. Please describe the encapsulation technology.

Chapter 5
PV Systems

In order to meet the needs of high power output, only solar cells are not enough, and it is necessary to form a photovoltaic power generation system together with other equipment.

5.1 PV systems and types

A solar photovoltaic power generation system generally consists of a solar cell array, a battery, a photovoltaic controller, and a DC/AC inverter (Figure 5.1). The photovoltaic power generation system converts the DC power generated by the solar cell into AC power that meets the load requirements or has the same voltage, frequency and phase as the grid, and transmits, stores and applies it.

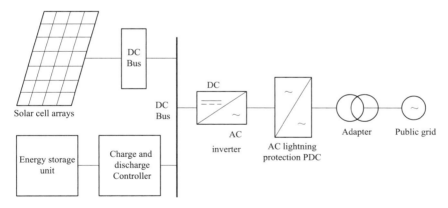

Figure 5.1 Photovoltaic power generation system structure

Among them, the solar cell is the most important part in the photovoltaic system, and its performance directly affects the efficiency of the whole system.

The battery is the energy storage device in the photovoltaic power generation system, which plays an important role in the smooth operation of the system. There are many types of batteries that meet the requirements of photovoltaic applications, including valve-regulated lead-acid batteries, all-vanadium flow batteries, sodium-sulfur batteries, and lithium iron phosphate batteries.

Chapter 5 PV Systems

The full name of the photovoltaic controller is solar charge and discharge controller, and it is an automatic control equipment, which controls the solar cell modules to charge the battery in an independent photovoltaic system, and controls the battery supplying power to the DC load or the AC load (via the inverter). Its role is to control the working state of the entire system.

The main function of the inverter is to invert the direct current output from the solar cell or battery into the alternating current.

There are a variety of electricity generation forms for ground – based photovoltaic systems. There are two main types of the PV system, namely, the stand – alone PV system and the grid – connected PV system. In addition, there are also hybrid system and small – scale power system.

> **Specialized Vocabulary**
> - PV system　光伏系统
> - grid　电网
> - stand – alone PV system　独立光伏发电系统
> - grid – connected PV system　并网光伏发电系统
> - hybrid system　混合系统
> - small – scale power system　小规模电源系统

5.2 Stand – alone PV systems

A stand – alone PV system is a power system powered by PV panels that are independent of the utility grid. This type system can use solar panels alone or in conjunction with a diesel (or natural gas or gasoline) generator or wind turbine. The design of a stand – alone power system based on photovoltaic is determined by location, climate, site characteristics and equipment to be used.

Figure 5.2 shows the schematic of a typical PV – based stand – alone power system. The independent PV power generation system mainly includes the following parts: photovoltaic array, photovoltaic controller, battery pack, off-grid inverter, monitoring system, and load. Over the past decade, many countries and regions have developed standards and/or guidelines for PV systems, and the applicable ones should always be understood and followed by designers and installers. Sometimes compliance is a condition for subsidies or other forms of financial support.

According to the characteristics of power load, independent PV power generation system can be divided into DC system, AC system and AC/DC hybrid system. There are two independent PV systems, namely, the direct coupled system without batteries and

5.2 Stand-alone PV systems

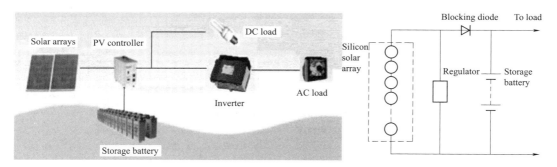

Figure 5.2　Simplified stand-alone PV power system

the stand-alone system with batteries (Figure 5.3). The basic model of a direct coupled system consists of solar panels directly connected to a DC load. Because there is no battery pack in this setup, energy is not stored, so it can only power common appliances such as fans and water pumps during the day. Maximum power point trackers (MPPTs) are commonly used to efficiently utilize solar energy, especially for electric loads such as positive displacement water pumps. Impedance matching is also considered as a design criterion for direct coupling systems. In standalone PV systems, the electricity generated by PV panels cannot always be used directly. Since the load requirement does not always equal the capacity of the solar panel, a battery pack is usually used. In a standalone PV system, the main functions of the battery are storage capacity and autonomy, storing excess energy and making it available when needed. Voltage and current stabilization provides stable current and voltage by eliminating transients. The surge current source can provide surge current for loads such as motors when needed.

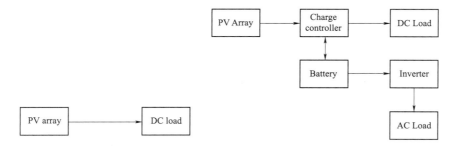

Figure 5.3　Direct-coupled PV system; Diagram of stand-alone PV system with battery storage powering DC and AC loads

Specialized Vocabulary

- direct-coupled system　直接耦合系统
- solar panel　太阳电池板
- direct current (DC)　直流
- maximum power point trackers (MPPTs)　最大功率点跟踪

Chapter 5 PV Systems

- AC load　交流负载
- inverter　逆变器

5.3 Grid connected PV systems

A grid-connected PV system is a power system powered by PV panels, which are connected to the public grid (Figure 5.4). The grid-connected PV power generation system consists of PV panels, MPPT, solar inverter, power conditioning unit and grid-connected equipment. Unlike stand-alone PV systems, these rarely have batteries. When conditions are right, grid-connected PV systems supply the excess power beyond the consumption of connected loads to the utility grid.

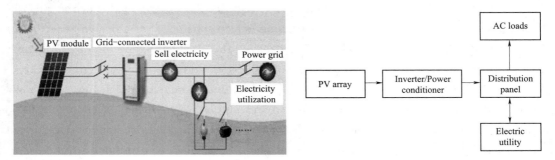

Figure 5.4 Grid-connected PV system

The capacity of residential grid-connected PV power system is relatively low, less than 10kW, which can meet the load of most users. They can channel excess power to the grid, which in this case acts as the system's battery. The feedback is through a meter to monitor the power transfer. This is called net metering. PV wattage may be below average consumption, in which case consumers will continue to buy grid energy, but in smaller quantities than before. If the PV wattage greatly exceeds the average consumption, the panels will produce far more energy than is needed. In this case, the excess power can yield revenue when it is sold to the grid. Under their agreement with the local grid energy company, consumers pay only the cost of the electricity consumed, which is less than the value of the generation. However, if there is significantly more electricity generated than consumed, this will be a negative number. In addition, in some cases, grid operators pay cash incentives to consumers. The connection of the PV system can only be done through an interconnection agreement between the consumer and the utility company. The protocol specifies in detail the various security standards to be followed during the connection.

Solar energy collected through PV solar panels, intended for delivery to a power grid, must be regulated or processed for use by grid-connected inverters. The inverter sits between the solar array and the grid, draws energy from both, either as a large stand-alone unit or as a set of small inverters, each connected to a separate solar panel. The inverter must monitor the grid voltage, waveform and frequency. One reason for monitoring is that if the grid is not operating or deviates too far from its nominal specifications, the inverter must not pass through any solar energy. According to safety rules (e. g. UL1741❶), the inverter connected to the malfunctioning power line is automatically disconnected, which vary by jurisdiction. Another reason for the inverter to monitor the grid is that when operating normally, the inverter must synchronize with the grid waveform and generate a voltage slightly higher than the grid itself so that energy flows smoothly outward from the solar array.

Grid-connected PV systems will reduce electricity bills as it is possible to sell surplus electricity generated to local power suppliers. Grid-connected PV systems are relatively easy to install because they do not require a battery system. Grid interconnection of PV system has the advantage of efficient use of power generation because it does not involve storage loss. PV systems are carbon negative over their lifetime because, any energy produced over and above that to build the panel initially offsets the need for burning fossil fuels. Even if the sun doesn't always shine, any installation can reasonably predict an average reduction in carbon consumption.

Specialized Vocabulary

- grid-connected photovoltaic power systems 并网光伏发电系统
- power conditioning units 功率调节单元，功率调节器
- residential grid-connected photovoltaic power systems 住宅光伏并网发电系统
- utility grid 公用电网
- fossil fuels 化石燃料
- carbon negative 碳负极
- solar array/PV array 太阳电池阵列/光伏阵列
- inverter/power conditioner 功率调节器
- distribution panel 配电盘

5.4 Hybrid power systems

Hybrid power systems combine PV with other multiple generation forms, usually a diesel

---

❶ Standard for Inverters, Converters, Controllers and Interconnection System Equipment for Use with Distributed Energy Resources.

Chapter 5 PV Systems

generator or biogas, to deliver the non-intermittent electric power. The other form of generation may be a type able to modulate power output as a function of demand. Furthermore, multiple forms of renewable energy, such as wind power, can be used. Photovoltaics can reduce the consumption of non-renewable fuels. Hybrid systems are most common on islands. Pellworm island in Germany and Kythnos island in Greece are famous examples (both are combined with wind energy). The Kythnos plant has reduced diesel consumption by 11.2 %.

It has also been shown that the deployment of a distributed network of PV + CHP (Combined Heat and Power) hybrid systems in the United States can improve the PV penetration limit. The temporal distribution of solar flux, electrical and heating requirements for representative U.S. single family residences were analyzed and the results clearly showed that hybridizing CHP with PV can enable additional PV deployment above what is possible with a conventional centralized electric generation system. This theory was again confirmed by numerical simulations using per second solar flux data to determine that it is possible to provide the necessary battery backup for such hybrid systems with relatively small and inexpensive battery systems. Moreover, large PV + CHP systems can also be used in institutional buildings, which again provide backup for intermittent PV and reduce CHP runtime. Diagram of PV hybrid system is shown in Figure 5.5.

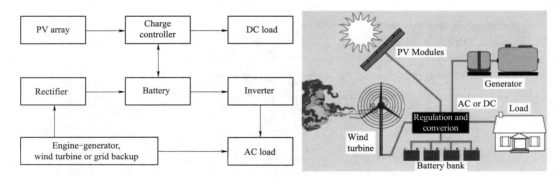

Figure 5.5 Diagram of photovoltaic hybrid system

Specialized Vocabulary

- diesel generator　柴油发电机
- biogas　生物气，沼气
- solar flux　太阳辐射通量
- centralized electric generation system　集中式发电系统
- combined heat and power　热电联供
- rectifier　整流器

5.4 Hybrid power systems

Reading Materials

1. *Design and development of distributed solar PV systems: Do the current tools work?*

2. *Review on sun tracking technology in solar PV system*

Reading Materials

Exercises and Discussion

1. Please describe the characteristics and differences among the stand-alone PV system, the grid-connected PV system and the hybrid system?

2. What are the applications for these three different PV systems?

Part II Solar Thermal Energy

Chapter 6

Solar thermal energy overview

6.1 The brief history of solar thermal energy

Solar thermal energy is a form of energy that generates heat from solar radiation. The earliest use of solar thermal energy can be traced back to the 7th century BC, but the use of solar energy as an energy source is only more than 300 years old. The solar thermal energy can be used for heating/cooling purposes for industrial, residential, power generation application and so on.

Great progress has been made in the utilization of solar thermal energy now. The first person to really use solar thermal energy in history was French scientist Solomon de Cox, who invented the world's first engine powered by solar heat energy in 1615. The machine uses solar thermal energy to heat air and make it work to pump water.

At the beginning of the 20th Century, solar energy research around the world is focused on solar power equipment, but concentrating light method is diversified. It is reported by some visionaries that traditional energy were rapidly declining after World War II. In order to solve this problem, people paid attention to the solar energy. During this time, solar energy research has made significant progress. The first solar power tower is designed in 1950, and the world's first solar – powered office building is designed by Frank. In 1952, the French National Research Center built a solar furnace in the eastern Pyrenees mountains. The basic theory of selective coating was proposed by Taber et al. in Israel in the first International Solar Thermal Science Conference in 1955. Black nickel and other selective coatings were developed to improve the collector efficient.

United States and Japanese government propose sunlight power generation/Sunshine project to develop solar thermal utilization technology in 1970s. The solar thermal energy research group and laboratories have emerged in university and research institutes in China. In 2021, United States announced funding program of Generation 3 Concentrating Solar Power Systems with 72 million $.

With the rapid development of technology and material, the research field of solar thermal energy is constantly expanding and deepening, and has made great achievements, such as in compound parabolic concentrator and vacuum collector tube. The solar thermal energy industry is preliminarily established with small scale compared with the traditional energy.

> **Specialized Vocabulary**
> - solar thermal energy   太阳热能
> - flat plate collector   平板集热器
> - traditional energy   传统能源
> - working medium   工作介质
> - solar radiation   太阳辐射
> - solar thermal power generation   太阳能热发电
> - vacuum collector tube   真空集热管

6.2 Solar thermal power generation system

Solar radiation heat energy collected by reflectors, concentrator and other heat collectors is gathered to the heat collector, which is used to heat molten salt or other heat transfer medium inside. Water is heated by heat transfer medium to high-temperature and pressure steam through the heat exchange device, and the electricity is generated by steam driving the turbine to actuate the generator. This kind of power generation technology through the conversion process of "light energy – heat energy – mechanical energy – electrical energy" is called concentrated solar power (CSP) generation technology. Compared with the fossil fuel power plant, the main difference is the heat source, which is from the abundant and clean solar thermal energy in CSP.

With the world's attention to energy issues, solar thermal power generation projects are gradually popular, and all countries have started or been ready for CSP projects. Figure 6.1 shows the tower solar power plant. As presented in Figure 6.2、Figure 6.3, in 2019, the global installed capacity of CSP increased by 381.6 MW, and total installed capacity is about 6,590 MW. Among them, the installed capacity in the Chinese market (200 MW) accounted for 52.41% of the world's new installed capacity.

Solar thermal power generation has incomparable preponderance such as the good matching with the power grid, the continuous and stable power generation capacity and peak-adjustment capacity, and the green production process of the power generation

Figure 6.1   Tower solar power plant

6.2 Solar thermal power generation system

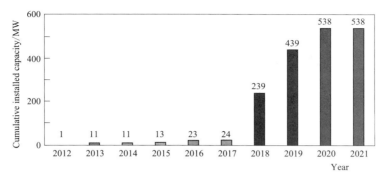

Figure 6.2 China's total installed solar thermal power capacity by the end of 2021

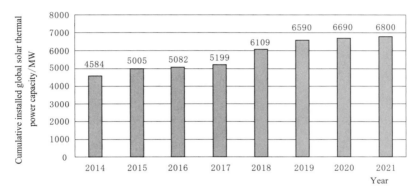

Figure 6.3 Cumulative installed global solar thermal power capacity

equipment. Therefore, solar thermal power generation has become a hot spot in the development and application of renewable energy in recent years. Countries have introduced corresponding economic support and incentive policies, and the global total installed capacity continues to rise, showing a scene of vigorous development. At present, solar thermal power demonstration stations have been widely built from 2016 in China. Molten salt and heat conduction oil are the mainly heat transfer and heat storage medium in solar thermal utilization.

In the future, development and application of the solar thermal power generation technology will toward the technical aspects of "high parameter, large capacity and continuous power generation". High parameters are high concentration ratio, thermoelectric conversion efficiency and operating temperature. Therefore, it is need to intensify research and development efforts in the key technologies and equipment including high-precision solar tracking control system, high-reflectivity and precision mirror and the solar thermoelectric conversion. To develop the CSP system, it is mainly to cut down the investment and construction cost, and gradually have the competitive ability comparable to the traditional thermal power generation. In general, solar thermal power generation tech-

Chapter 6 Solar thermal energy overview

nology will rapidly develop in the direction of low cost and large scale, and will play a pivotal role in the future energy structure.

Specialized Vocabulary

- reflectors 反射镜
- concentrator 集热器
- concentrated solar power generation technology 聚光太阳能发电技术
- concentrating solar power 聚光太阳能电站
- renewable energy 可再生能源
- concentration ratio 聚光比
- operating temperature 工作温度
- heat conduction oil 导热油
- molten salt 熔盐
- solar thermal utilization 光热利用
- incomparable preponderance 无与伦比的优势

Reading Materials

1. *Mapping the development of various solar thermal technologies with hype cycle analysis*
2. *What is the history of solar energy and when were solar panels invented?*

Reading Materials

Exercises and Discussion

1. What applications solar thermal energy can be used for?
2. How old is the history of solar thermal utilization?
3. Which aspect of solar thermal utilization do you think will be widely used in the future? Why?
4. What are the prospects for solar thermal utilization?

Chapter 7

Solar thermal power generation system

7.1 Brief introduction of solar thermal power generation

In 1866, the first application of solar thermal power is documented, which is using the parabolic trough device to generate steam to run the engine. The first solar thermal power plant is built and operated in 1968. The Solar One and Solar Two tower solar thermal power plant are built in 1982 and 1996 in the USA. Thirty-three concentrated solar power plants are under construction in Spain. Meanwhile, the solar thermal power technology in China, Egypt, France, Morocco, Algeria and other countries are quickly developed in recent years.

The main principle of solar thermal power generation is using the optical system to gather solar radiation energy, and steam is generated by collected solar thermal energy heating water, steam drive the turbine generator to generate electricity. The operation of the whole system can be divided into three steps: heat collection, heat absorption of the thermal medium, drive the engine to generate electricity. Compared with photovoltaic power generation, it has the advantages of high efficiency, compact structure and solar thermal energy storage. The solar thermal power generation system can be divided into four types: parabolic trough solar thermal power system, tower solar thermal power system, disk solar thermal power system and linear Fresnel solar thermal power system. At present, the parabolic trough and tower solar thermal power generation system are the most mature and have reached the level of commercialization.

Solar thermal power generation needs higher concentration rate and large-scale thermal energy storage. Therefore, solar thermal power plant is built in the wide area with abundant sunlight. For heat conduction and thermal storage medium, the molten salt is popularly used due to the low cost, long service life performance. In terms of energy conversion, photovoltaic power generation is from light energy to electric energy, while solar thermal power generation is from light energy to heat energy to electric energy. Therefore, the solar thermal power generation system is more complicated compared with photovoltaic power generation system. In general, solar thermal power generation technology is complicated and has been constantly improving.

Chapter 7 Solar thermal power generation system

Specialized Vocabulary

- optical system 光学系统
- compact structure 紧凑结构
- parabolic trough solar thermal power system 槽式太阳能热发电系统
- tower solar thermal power system 塔式太阳能热发电系统
- disk solar thermal power system 碟式太阳能热发电系统
- linear Fresnel solar thermal power system 线性菲涅尔式太阳能热发电系统
- concentrated solar power plant 聚光型太阳能光热发电

7.2 Parabolic trough solar thermal power system

As shown in Figure 7.1, parabolic trough solar thermal power system uses a plurality of trough parabolic reflectors to focus the solar energy on a line, and a tubular collector is installed on the focusing line to absorb the concentrated solar energy to generate electricity. The system generally consists of a light collector, a heat storage device, a thermal generator or/and an auxiliary energy device (such as a boiler). The concentrator and heat collector system are the core of the system,

Figure 7.1 Trough solar collector

which consists of a concentrator, a receiver and a tracking device. Its concentrating ratio is from 10 to 100, the heat transfer medium temperature in tubular collector can reach more than 400℃. Parabolic trough solar thermal power system is widely used around the world.

The United States, as a leader in the solar thermal power generation technology, has made outstanding achievements. Since 1945, LUZ has successively built nine solar trough power stations in the California desert. In 2007, America's trough power system went online in the state of Nevada, and 12 new ones were added. By 2012, about 3,100 MW of trough power systems were connected to the grid in the United States. Currently, many countries have initiated plans to build trough photovoltaic thermal power stations, as shown is Table 7.1.

Table 7.1   Part of parabolic trough solar thermal power project

Country	Year	Project name	Working medium	Scale/MW
United States	1989	Solar Electric Generating Station Ⅷ	Biphenyl/Diphenyl oxide: Therminol VP-1	80
Spain	2011	Andasol 3	Thermal Oil	50
Spain	2011	Arcosol50 (Valle I)	Biphenyl/Diphenyl oxide	50

7.3 Tower solar thermal power system

continued

Country	Year	Project name	Working medium	Scale/MW
India	2013	Godawari Solar Project	Biphenyl/Diphenyl oxide: Dowtherm A	50
India	2013	KVK Energy Solar Project	Synthetic Oil	100
Morocco	2014	Airlight Energy Ait – Baha Pilot Plant	Air	3
United States	2014	Genesis Solar Energy Project	Biphenyl/Diphenyl oxide: Therminol VP – 1	250
South Africa	2018	Ilanga I	Thermal Oil	100
China	2018	CGN Delingha – 50MW Trough	Biphenyl/Diphenyl oxide: Therminol VP – 1	50
China	2020	CSNP Urat – 100MW Trough	Thermal oil	100

Although the application of trough solar collector in China started late, with the continuous research of domestic experts and scholars, great breakthroughs have been made in recent years. The Changchun Institute of Electrical Engineering of the Chinese Academy of Sciences, together with the Institute of optics mechanics of the Chinese Academy of Sciences, Southeast University and other institutions, has conducted comprehensive research on solar thermal power generation technology, focusing on the research of reflectors, heat absorption tubes, control systems and key materials, and has made great progress. In 2011, a 180 kW trough power station was successfully connected to the grid in Turpan, Xinjiang. In January 2020, the largest national demonstration project of 100 – MW level trough thermal power generation – Urad Middle Banner thermal oil Tank 100,000 kW, 10 hour energy storage thermal power generation project was successfully turned over once and entered the trial operation stage. The project is the largest parabolic trough solar thermal project in China. The installed capacity is 350 million kWh per year. Compared with traditional thermal power plant of the same size, 120,000 tons of standard coal can be saved per year.

Specialized Vocabulary

- tubular collector 管状集热器
- auxiliary energy device 辅助能源装置
- demonstration project 示范工程
- trial operation 试运行
- heat transfer medium 传热介质
- trough solar collector 槽式聚光器

7.3 Tower solar thermal power system

Figure 7.2 shows the tower concentrated solar thermal power station, the tower solar thermal power system was firstly put forward in the 1950s. The heat transfer medi-

Chapter 7 Solar thermal power generation system

Figure 7.2 Tower solar thermal power station

um in tower solar thermal power system can be molten salt, water, liquid sodium, heat conduction oil. The concentrating ratio can reach from 300 to 1500, and the operating temperature in tower solar thermal power system can reach 1500 ℃.

The basic form of tower solar thermal power generation system is to use a cluster of heliostats independently tracking the sun to concentrate sunlight on a fixed receiver at the top of the tower, which generates high temperature heat transfer medium, heats steam to drive a conventional generator set to generate electricity. This kind of power generation does not need the traditional energy, and the power supply comes entirely from the high temperature heat transfer medium produced by solar radiation in the heat collecting system. The tower thermal power generation system mainly consists of heliostat array, high tower, heat receiver, heat transfer fluid, heat exchange components, heat storage system, control system, steam turbine and power generation system.

Tower solar thermal power generation is characterized by the high temperature molten salt to store solar thermal energy, the high concentration ratio, easy to achieve a higher working temperature, and the relatively smaller receiver heat dissipation area, which can get a higher solar - thermal conversion efficiency. The operation parameters of tower solar thermal power station are basically same as the conventional thermal power station, so it not only has higher thermal efficiency, but also is easy to obtain auxiliary equipment.

As listed in Table 7.2, tower solar thermal power generation technology was born earlier. In the late 19th century, Germany, Italy, France and other nine countries jointly built a power station, whose installed capacity reached 1MW, is the world's first tower solar thermal power station. In 1980s and 1990s, the United States built Solar One, which generates 10MW of electricity, and later successfully built Solar Two, which can run a steam engine at full capacity for three hours under maximum storage conditions. Tower solar thermal power generation technology has gradually become an important subject of academic research.

Table 7.2 Part of tower solar thermal power project

Country	Start year	Project name	Working medium	Scale/MW
India	2011	ACME Solar Tower	Water	2.5
Spain	2011	Gemasolar Thermosolar Plant/Solar TRES	Molten salt	20
United States	2015	Crescent Dunes Solar Energy Project	Molten Salt	110

continued

Country	Start year	Project name	Working medium	Scale/MW
Australia	2017	Jemalong Solar Thermal Station	Liquid sodium	1.1
Israel	2019	Ashalim Plot B/Megalim	Water	121
China	2019	Power China Qinghai Gonghe-50 MW Tower	Molten salt	50
China	2021	Shouhang Yumen 100 MW Tower	Molten salt	100

Specialized Vocabulary

- heat transfer 传热
- liquid sodium 液态钠
- cluster 群,束
- heliostats 定日镜
- conventional energy 传统能源
- operation parameter 运行参数
- thermal efficiency 热效率

7.4 Dish solar thermal power system

As shown in Figure 7.3, dish solar thermal power system uses the two-axis tracking, the rotating parabolic mirror, pointing to the incident solar radiation, and the heat absorber absorbs this portion of the radiant energy and converts it into heat energy, heating medium to drive the engine (such as gas turbine engine, Stirling engine or other type), so as to convert heat energy into electricity.

Compared with the previous two kinds of solar thermal power generation technology, the development history of the disk solar thermal power generation technology is

Figure 7.3　Dish solar thermal power system

a little short, but its development speed is very fast. As early as 1878, a small disc installation was installed in Paris, France. The original purpose was to use the concentrated solar energy to power boilers. Its feasibility was further demonstrated in the early 20th century when California successfully built the first solar concentrator system. Eighty years later, California completed the Dish Stirling solar thermal system, which greatly improved the efficiency of power generation by nearly 25 kW and achieved a conversion efficiency of 29%. In recent years, the United States built more powerful and efficient solar dish systems (Table 7.3).

Table 7.3　　　　　　　Part of dish solar thermal power project

Country	Start year	Project name	Working medium	Scale/MW
United States	2010	Maricopa Solar Project	—	1.5
	2012	Tooele Army Depot	Helium	1.5

Specialized Vocabulary

- two-axis tracking 双轴追踪
- rotating parabolic mirror 旋转抛物面反射镜
- Stirling engine 斯特林发动机
- gas turbine engine 燃气轮机
- feasibility 可行性
- conversion efficiency 转换效率

7.5　Linear Fresnel solar thermal power system

Linear Fresnel Reflector (LFR) solar thermal power system consists mainly of slightly curved mirrors that focus the solar energy on a fixed receiver (Figure 7.4). LFR mirrors provide uniaxial tracking, focusing incoming solar radiation energy on a small aperture receiver in sync with the motion of the solar disk. In order to obtain higher concentrated ratio and improve thermal efficiency, a secondary mirror with a smaller radius of curvature is used to form a receiving cavity around the absorption tube. The concentration ratio of linear Fresnel concentrator is between 35 and 100, and the fluid temperature in the tube can reach 565 ℃. A linear Fresnel concentrator can be conceived as a piece wide discretization of the reflector of a trough system, so that its mirrors need not remain parabolic and can be in the same plane. Compared with the trough, the linear Fresnel technology has further reduced construction and operating costs and has great commercial prospects. Linear Fresnel solar thermal power system has the following characteristics.

Figure 7.4　Linear Fresnel mirror field

(1) The concentrating ratio can usually reach 10~80, the average annual concentrating efficiency can reach about 10%, and the steam temperature can reach within the

range of 250~500 ℃.

(2) The main mirror is a combination of the flat and slightly curved strip mirror structure, and the secondary mirror and parabolic trough mirror have a lot in common, so the related production process is relatively perfect.

(3) The design of the main mirror is more scientific, for it can be regular, compact layout of a large area to increase the utilization rate of land, and the mirror of the device is installed close to the ground to minimize the wind resistance, so its site selection is not of much restriction.

(4) The fixed position of the collector does not change with other movements, which greatly reduces the sealing and connection problems as well as the cost.

(5) Because the flat mirror is used, it is easier to clean with less water used, and the cost is reduced.

The emergence of the linear Fresnel concentrator began in 1957. Baum et al. studied and analyzed such large biaxial tracking systems, and proposed for the first time the idea of splitting a large mirror into several small Fresnel discrete mirrors to improve the adaptability of the device.

In the 1990s, a linear Fresnel concentrating system with secondary reflectors was designed by Paz in Israel to improve the concentrating ratio and maintain operating efficiency at high temperature. At present, in order to improve low energy density and the uneven distribution of heat flux condition, researchers have designed many secondary concentrators with different structures (Table 7.4). In order to improve the concentrating performance, some other parameters affecting the concentrating performance are also studied. Many researchers have carried out experiments on multiple prototypes, studied the performance characteristics of linear Fresnel reflector solar thermal power under different optical and geometric parameters, reviewed the technical characteristics under different design concepts, and prospected the development prospects of the linear Fresnel solar thermal power technology.

Table 7.4   Some linear Fresnel solar thermal power projects

Country	Year	Project name	Working medium	Scale/MW
France	2012	Augustin Fresnel 1	Water	0.3
Australia	2012	Liddell Power Station	Water	3
Spain	2012	Puerto Errado 2 Thermosolar Power Plant	Water	30
India	2014	Dhursar	Water	125
Italy	2014	Rende – CSP Plant	Diathermic oil	1
China	2018	Huaqiang TeraSolar 15MW Fresnel	Water	15
China	2019	Lanzhou Dacheng Dunhuang – 50MW Fresnel	Molten salt	50

Chapter 7   Solar thermal power generation system

Specialized Vocabulary

- aperture receiver　孔径接收器
- receiving cavity　接收腔
- discretization　离散化
- wind resistance　风阻
- secondary reflector　二次反射镜
- in sync　同时
- utilization rate　利用率
- mirror field　镜场
- energy density　能量密度
- uneven distribution　不均匀分布
- heat flux　热流密度
- prototype　模型

Reading Materials

1. *Concentrated solar power* (CSP), *explained*
2. *On the use of thermal energy storage in solar-aided power generation systems*

Reading Materials

Exercises and Discussion

1. What do you think of the solar thermal power generation situation in China?

2. What is the operating principle of the solar thermal power generation system?

3. What is the change of energy form in the operation of the solar thermal power generation system?

4. What are the heat transfer working materials of these four solar thermal power generation systems?

5. Which solar thermal power system do you think will be widely used in the future? Why?

Chapter 8

Solar heating and cooling system

8.1 Solar heating system

8.1.1 Active solar heating system

Depending on whether external driving force is needed, solar heating systems can be divided into active solar heating system and passive solar heating system. Active solar heating system including solar collector, fan or pump, pipe, heat storage device, and indoor cooling terminal, is a forced circulation solar heating system. It transmits the heat transfer working medium (water or air) to the accumulator or the room to be heated through the solar collector. The system is generally equipped with auxiliary heat source device to ensure the heating effect. The most critical parts are solar collector, the auxiliary heat source and heat storage device.

8.1.2 The core components

Solar collector is the core part of the solar heating system, which mainly adopts flat plate collector, vacuum tube collector and compound parabolic collector. Flat plate collector and vacuum tube collector are mostly applied in China. At present, the study on the collector in active solar heating system mainly focuses on the integrated design of the building and the collector, so as to realize the building heating and ensure the elegant appearance of the building, the improved collector to increase efficiency and heat collection temperature, optimization of the collector area of the system and the best inclination angle of the collector, to name but a few.

8.1.3 Auxiliary heat source

Since solar radiation is intermittent and unstable, the heating demand in a certain period is characterized by continuity and stability. In order to ensure the stability and reliability of solar heating system and meet the requirements of the heating comfort, auxiliary heat sources must be set up. The current forms of the solar energy heating system of the auxiliary heat source include electric heaters, oil-fired boilers, gas boilers, coal boilers, biomass boilers, urban and industrial waste heat, etc. In recent years, the auxiliary heat source is popularly using the ground source heat pump, and the joint of the solar-ground

Chapter 8 Solar heating and cooling system

source heat pump system. For the setting of the auxiliary heat source, we should comprehensively consider the local solar resource conditions, the conventional energy supply and the heating load, and optimize the auxiliary heat source based on the analysis of economy (such as the annual cost and the initial investment), the system performance and energy saving, and design the solar guarantee rate of the system reasonably.

8.1.4 Heat storage device

The periodicity and instability of the solar radiation cause the time mismatch between the heating demand and the solar energy. In order to solve this problem and fully exploit the potential of solar energy resources, the solar heating system often needs to be equipped with heat storage devices. Based on the heat storage time, the heat storage device can be divided into short-term and long-term heat storage. Heat storage device can be divided into sensible, latent and chemical heat storage according to heat storage mechanism. For short-term heat storage, the main heat storage methods are heat storage tank and phase change heat storage; for long-term heat storage, especially for cross-season heat storage, sensible heat storage devices are more commonly used, including hot water heat storage, buried pipe heat storage, aquifer heat storage and gravel-water heat storage.

Specialized Vocabulary

- forced circulation　强制循环
- inclination angle　倾斜角
- oil-fired boiler　燃油锅炉
- biomass boiler　生物质锅炉
- heat pump　热泵
- ground source heat pump　地源热泵
- initial investment　初投资
- annual cost　费用年值
- short-term heat storage　短期蓄热
- long-term heat storage　长期蓄热
- sensible heat storage　显热蓄热
- latent heat storage　潜热蓄热
- chemical heat storage　化学蓄热
- heat storage tank　蓄热水箱
- phase change heat storage　相变蓄热
- hot water heat storage　热水蓄热
- buried pipe heat storage　地埋管蓄热
- aquifer heat storage　含水层蓄热
- gravel-water heat storage　砾石-水蓄热

8.2 Passive solar heating system

Passive solar heating system is commonly known as passive solar house. Through the reasonable layout of the building orientation and the surrounding environment, the proper choice of building materials, structures, internal space and the external form, the building can fully absorb and store solar radiation energy in winter, so as to achieve building heating without using additional heating equipment. The passive solar heating system generally does not require the use of mechanical equipment and power, which is the biggest feature that distinguishes it from other heating systems.

The passive solar heating system mainly uses the principle of the greenhouse effect, based on heat collection, heat storage and lighting components for the heat transfer and utilization. A south window is the most basic component, which can fully absorb the solar radiation. The ground, wall and other envelope structures usually as heat collection and heat storage components, can be used as a part of the envelope structure, at the same time to achieve the utilization of solar thermal energy and play a multi-functional role of the structure. It has simple structure, low cost, and easy to operate, maintain and manage, but with the change of the solar radiation, the heat collection efficiency is relatively low, the change of heat in the heating space fluctuates greatly, and the heating quality is poor. According to the way of the solar energy utilization, the passive solar heating technology can be divided into direct-gain type, thermal storage wall type, additional sunspace type and combination type, which are also the four most commonly used technology types in life.

Specialized Vocabulary

- surrounding environment 周围环境
- greenhouse effect 温室效应
- passive solar house 被动式太阳房
- envelope structure 围护结构
- heat collection efficiency 集热效率
- direct-gain type 直接受益式
- thermal storage wall type 集热蓄热墙式
- additional sunspace type 附加阳光间式
- combination type 组合式

8.2.1 Direct-gain type

Because of its simple structure, the direct-gain type is the most commonly applied and the earliest passive solar heating method. In order to facilitate the direct-gain window

Chapter 8 Solar heating and cooling system

and other transparent envelopes that directly receive solar radiation, the system is set in the south and selected with the high transmission of glass. Most of the sunlight directly irradiates through the transparent envelope to the indoor ground and walls, making them fully absorb heat and raise the temperature, a small amount of sunlight is reflected to other surfaces in the room, where it is absorbed and reflected again. The quantity of the heat absorbed by the surface inside the envelope structure is introduced indoors with convection, radiation and the means that heat conducts, and waits for the indoor temperature to drop or a cloudy day to release, maintaining the indoor temperature at a higher level.

In order to further increase the average indoor temperature of passive solar heating buildings in winter, a larger window-wall ratio is often used to increase daytime heat collection, but it may also cause overheating. So, thermal insulation curtains are usually equipped to reduce daytime heat absorption and nighttime heat dissipation. In particular, in order to further improve the heat storage performance and reduce the heat loss, the outer protective structure should have large thermal resistance, and there are heavy materials such as brick, concrete and adobe with good heat storage and release performance in the room. These materials can absorb enough heat during the day time, and release abundant heat at night when the outdoor temperature drops, so as to achieve a better heating effect (Figure 8.1).

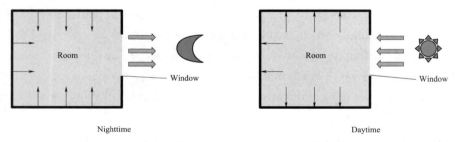

Figure 8.1 Direct-gain type at nighttime, direct-gain type at daytime

Specialized Vocabulary

- convection 对流
- radiation 辐射
- heat conducts 热传导
- window-wall ratio 窗墙比
- overheating 过热
- indoor temperature 室内温度
- thermal insulation curtains 保温窗帘
- thermal resistance 热阻
- brick 砖
- concrete 混凝土
- adobe 黏土

8.2 Passive solar heating system

8.2.2 Thermal storage wall type

The thermal storage wall is composed of glass covers, bricks, concrete, adobe and other heat storage materials with good heat storage performance. A heavy structural wall is set behind a south - facing transparent glass window as a heat collection device, and heat absorption materials with a high absorption rate are coated on the wall surface to promote the absorption of the sunlight. The heat storage wall has better heat storage capacity, and its application principle is as follows: When the sunlight passes through the southward glass, part of the heat goes through the way of heat conduction from the outer surface into the inner surface of the wall, and then into the room through convection and radiation, while another part of the heat will also go from the outer surface of the wall and the glass interlayer of the air convection heat transfer, and then through convection into the room.

There are many forms of thermal storage walls. According to whether the wall has ventilation holes, they can be divided into those with ventilation outlets and those without. The non - vent type mainly uses the wall heat conduction as the heat transfer mode to provide the indoor heating. The heat collected by the vent wall during the day passes through the vent on the wall and the indoor heat exchange in the form of convection. The vent is closed at night, and the heat transfers to the indoor heating through the heat conduction. The thermal storage wall can be divided into lattice type, solid type, water wall type and other forms according to forms and structures. The solid type absorbs solar radiation with a glass cover covering the south wall. The main difference between the lattice type and the solid one is that the walls are covered with ventilation holes and equipped with insulation boards and baffles. During the day, the insulation panels open, the baffles close to allow the walls to store heat, and at night the insulation panels close to release heat into the room. The heat collection and storage wall of the water wall type is located in the south, and the water is placed in one wall of the room. Its operation management is very complicated, and it is not often used. The solid heat collection wall with ventilation outlets is the most common form of heat collection and storage walls. It is convenient to build and easy to use, and its principle is shown in Figure 8.2.

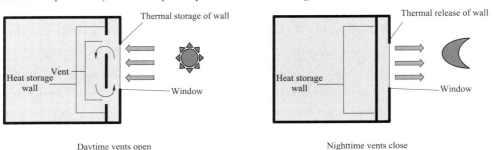

Figure 8.2 Thermal storage wall type at daytime; thermal storage wall type at nighttime

Chapter 8 Solar heating and cooling system

Specialized Vocabulary
- glass cover 玻璃罩
- heat storage capacity 蓄热能力
- ventilation hole 通风孔
- insulation panel 隔热板
- lattice type 花格式
- solid type 实体式
- water wall type 水墙式

8.2.3 Additional sunspace type

The common wall for heat collection separates the additional space on the south side from the main room to form a basic sunspace–attaching structure. During the day time, the sunlight penetrating through the transparent window is fully absorbed by the ground and the common wall of the additional space. At night, the additional space can be used as a buffer to weaken the heat dissipation of the main room to the outside, so as to maintain a moderate environment of the main room. At the same time, the additional space as a buffer can effectively reduce the room temperature fluctuation and the glare of the main room, and maintain better comfort. The heating principle of the sunspace–attaching technology is shown in the figure below.

The passive solar heating technology with the additional sunlight combines the principle of the direct benefit type and the thermal storage wall type. With the application of this technology, the heating process can be divided into two stages: daytime and nighttime. During the day, the sun room is in a state of heat collection, absorbing solar radiation heat to heat the air, the wall and the ground in the sun room. When the temperature in the additional space is higher than that in the main room, the doors and windows (ventilation holes) are opened, and the air forms a natural circulation between the additional space and the adjacent room under the action of hot pressure. When the indoor temperature rises, the envelope structure of the room will also store heat to maintain the indoor temperature at a higher level. Due to the lack of solar radiation at night, the outdoor temperature is low, and the sunlight will heat the outdoor. At this time, the doors and windows of the shared wall (ventilation holes) are closed, and the sunshine room, as an insulating air compartment, plays a buffer role in the overall heat transfer process of the room to further reduce the room temperature fluctuation and maintain the stability of the room thermal environment (Figure 8.3).

8.2 Passive solar heating system

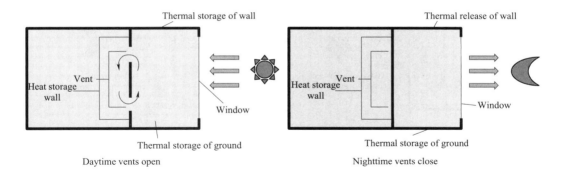

Figure 8.3 additional sunspace type at daytime and nighttime

Specialized Vocabulary

- additional sunspace type 附加阳光房式
- buffer 缓冲区
- heat dissipation 散热
- natural circulation 自然循环
- temperature fluctuation 温度波动
- glare 眩光
- penetrate 穿透

8.2.4 Combination type

The combined passive solar heating technology uses a variety of heat collection components. The additional sunspace technology is usually used in conjunction with windows, doors or other vents to satisfy both daylighting and heating requirements. The application of the direct-gain heating technology is susceptible to the impact of outdoor temperature changes, and sometimes sunspaces-attaching will also affect the indoor lighting effect. The two technologies are often combined in cold areas and rural areas, which can make full use of solar energy and improve the energy saving efficiency. In order to ensure the needs of lighting and ventilation and increase the total amount of the heat supply, the direct-gain window and other types of heat collection components such as heat collection walls are usually combined. In the daytime, the heat collection wall with ventilation outlet can provide heat supply through air convection, and at night, the wall conducts heat to ensure constant heat supply throughout the day. The use of the heat collection and storage wall without vents and the direct-gain window can also ensure the uniform heating in different periods of day and night (Figure 8.4).

Chapter 8 Solar heating and cooling system

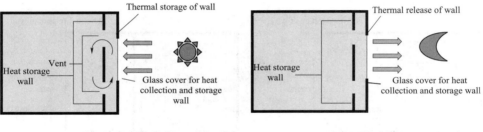

Figure 8.4 combination type at daytime and at nighttime

Specialized Vocabulary

- combination type 组合式
- in conjunction with 共同，配合
- air convection 空气对流
- rural area 郊区，农村地区
- rural areas 农村地区
- ventilation outlet 出风口

8.3 Solar cooling system

Generally speaking, summer is the time of the year when the solar radiation is the highest and the demand for air conditioning is the highest. It can be said that the demand for refrigeration is basically consistent with the intensity of solar thermal energy. If solar energy can be used to achieve refrigeration, the high solar radiation intensity in summer will benefit the solar driven air-conditioning system to produce more cold. Compared with other solar energy systems, the solar refrigeration system has better seasonal adaptability. The solar refrigeration system includes: solar photovoltaic system driven by the traditional compressor refrigeration, solar driven desiccant evaporative refrigeration, solar steam injection refrigeration, solar solid adsorption refrigeration, and solar drying cooling refrigeration.

Specialized Vocabulary

- solar driven air-conditioning system 太阳能驱动空调系统
- refrigeration 制冷
- seasonal 季节的
- adaptability 适应性
- desiccant 干燥剂

8.3 Solar cooling system

- adsorption 吸附
- solar driven desiccant evaporative refrigeration 太阳能驱动蒸发式制冷
- solar solid adsorption refrigeration 太阳能固体吸附式制冷
- solar drying cooling refrigeration 太阳能干燥冷却制冷
- solar steam injection refrigeration 太阳能蒸汽喷射制冷

8.3.1 Solar absorption refrigeration

The absorption refrigeration uses liquid refrigerant to vaporize and absorb heat under low pressure and temperature to cool the environment. The solar thermal energy in the refrigeration system is used to produce the high temperature hot water by the solar collector as the heat source of the refrigerator generator. A refrigerator consists mainly of four heat exchangers (generator, evaporator, condenser and absorber). The working medium of the system is generally composed of two substances with large difference in the boiling point. The working medium usually used is ammonia – water (Figure 8.5).

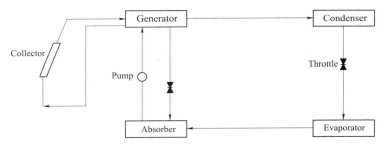

Figure 8.5　Solar absorption refrigeration

Specialized Vocabulary

- ammonia 氨
- solar absorption refrigeration 太阳能吸收式制冷
- boiling point 沸点
- ammonia – water 氨水
- absorption refrigeration 吸收式制冷

8.3.2 Solar driven desiccant evaporative refrigeration

Solar driven desiccant evaporative refrigeration and absorption refrigeration thermal cycle are similar, and the difference is that the former uses the solid adsorbent instead of the absorption refrigeration cycle absorbent. The system structure and operation are relatively simple, but the cycle is long, the refrigeration power is relatively small, and the refrigeration coefficient is low, which become the development disadvantages (Figure 8.6).

Chapter 8 Solar heating and cooling system

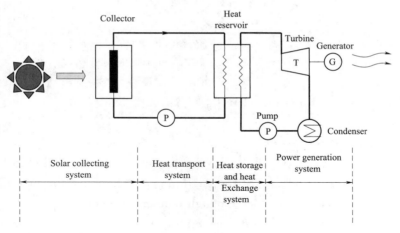

Figure 8.6 Solar driven desiccant evaporative refrigeration

8.3.3 Solar steam jet refrigeration

Jet refrigeration is the use of the steam from the ejector caused by the surrounding low pressure to drive the refrigerant evaporation, so as to reach the refrigeration effect. Solar thermal energy provides heat to the system to evaporate the steam. In theory, the steam jet refrigeration can be applied to general refrigerants, such as ammonia. Currently, water is commonly used, but the injection refrigeration coefficient is very low, which is still in the research stage and has not been widely used.

Specialized Vocabulary

- refrigeration power   制冷功率
- refrigeration coefficient   制冷系数
- jet refrigeration   射流制冷
- ejector   喷射器

Reading Materials

1. *Advances on multi-purpose solar adsorption systems for domestic refrigeration and water heating*

2. *Exergo-environmental cost optimization of a solar-based cooling and heating system considering equivalent emissions of life-cycle chain*

Reading Materials

Exercises and Discussion

1. What is the main basis for distinguishing active and passive solar heating systems?

8.3 Solar cooling system

    2. Which do you think should be used in large buildings, active or passive solar heating systems?

    3. What is the principle of the passive solar heating system?

    4. What are the main forms of passive solar heating systems?

    5. Briefly introduce the working principle of the solar refrigeration.

Chapter 9

Non – concentrating solar collector

Non – concentrating solar collector mainly includes vacuum tube collector and flat plate collector. The flat plate solar collector receives solar radiation thermal energy and transfers heat to the heat transfer working medium. The structure of the absorber is basically a flat plate shape. The vacuum tube collector is developed on the basis of the flat plate collector. The vacuum tube can effectively restrain the conduction and convective heat loss to the ambient enviroment.

9.1 Flat plate collector

The working principle of the flat plate collector is: sunlight shines on the heat collecting plate core through the glass cover plate, and the heat collecting plate core converts solar thermal energy and transfers it to the working medium in the flow channel, thus completing the conversion process from solar energy to thermal energy. The external insulation material is to reduce heat loss. Moreover, the selective absorption coating has low infrared emissivity, which can significantly reduce the radiant heat loss of the heat absorbing plate.

At present, the commercial flat plate collector technology is quite mature, and the current national standard stipulates that its intercept thermal efficiency should not be less than 72%, and its heat loss coefficient should not be higher than 6.0 W/ (m^2 · K). The heat loss coefficient of the high – performance flat plate collector can be lower than 4.0 W/ (m^2 · K). The current research focuses on reducing the cost of the collector by using new materials and processes, reducing the heat loss coefficient by means of the structure or material improvement, improving the intercept thermal efficiency and the frost resistance of the collector.

Commonly used medium – temperature flat plate collectors are: cellular flat plate collectors and CPC flat plate collectors. The operating temperature of the CPC flat plate collector exceeds 80℃, and that of the cellular flat plate collector exceeds 100～120℃. Medium – temperature flat plate collectors can greatly reduce the heat loss of the collector. High temperature flat plate collector with the reflector can collect more sunlight on the collector, so as to improve its thermal efficiency. Its operating temperature is generally up to 250℃. The main types of high-temperature flat plate collectors are vacuum flat plate collector and special-shaped flat plate collectors. In addition to the above common flat plate collectors, there are new collectors such as special-shaped flow channel flat plate collector, plastic film-insulated flat plate collector, serpentine flow channel flat plate collector and flat plate air collector.

Specialized Vocabulary

- non-concentrating solar collector 非聚光型太阳能集热器
- selective absorption coating 选择性吸收涂层
- frost resistance 抗寒性
- plastic film-insulated flat plate collector 塑料薄膜绝缘平板集热器
- serpentine flow channel flat plate collector 蛇形流槽平板收集器
- cellular flat plate collector 蜂窝式平板集热器
- flat plate collector 平板集热器
- operating temperature 工作温度

9.1.1 Cellular flat plate collector

Israel TIGI high-performance cellular flat plate collector: the transparent insulation allows sunlight to enter, and the dark surface absorbs solar radiation energy and converts it into thermal energy; the transparent insulation minimizes heat loss by reducing thermal convection and adverse radiation. The heat production per unit collector array area is significantly improved. The core of honeycomb collector is polymer in transparent insulating layer. Sunlight can enter the transparent insulation layer, thus allowing solar energy to enter the collector; the residual air cannot circulate in the transparent insulation layer, thus greatly reducing the heat convection loss; The polymer also blocks the infrared radiation, further reducing the heat loss. The glass cover plate has a light transmittance of 96% and can withstand pressure of 8bar and operating temperature above 110℃. The structure is shown in Figure 9.1.

Specialized Vocabulary

- polymer 聚合物
- honeycomb 蜂窝型
- insulating layer 绝缘层
- residual air 余气
- infrared radiation 红外辐射
- light transmittance 透光率

Chapter 9 Non-concentrating solar collector

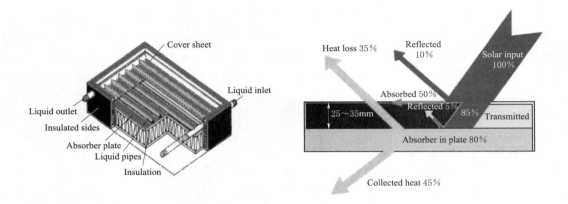

Figure 9.1 The structure of the cellular flat plate collector

9.1.2 CPC flat plate collector

A CPC flat plate collector refers to a flat plate collector with the CPC reflector. It can make full use of the sunlight, especially in spring and autumn. A trough reflector will reflect the incident sunlight to the vertical built-in heat absorption belt, and it does not directly contact with the external environment, so its service life is long. The heat absorption device adopts the copper selective absorption film, which can absorb the direct sunlight radiation and the scattered radiation with an absorption rate up to 95%.

The collector cover plate is made of solar toughened glass, with high light transmittance, pressure resistance and corrosion resistance (Figure 9.2). The operating temperature is about 120℃, and the collector's service life is up to 25 years. The enclosed collector package prevents pollutants and water vapor from infiltrating into the collector, which can resist extreme weather and temperature and slow down the aging of the collector. The collector seal will result in a negative pressure of 0.3bar inside the collector, which is just enough to be offset by the reflector support.

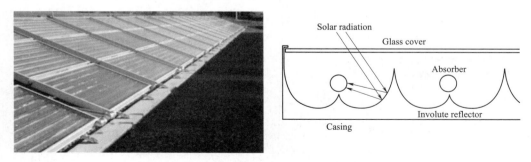

Figure 9.2 CPC flat plate collector

9.1 Flat plate collector

> **Specialized Vocabulary**
> - selective absorption film 选择性吸收膜
> - scattered radiation 散射辐射
> - toughened glass 钢化玻璃
> - light transmittance 透光率
> - corrosion 腐蚀
> - pollutants 污染物
> - infiltrating 渗透

9.1.3 Vacuum flat plate collector

The thermal utilization temperature of the TVP vacuum flat plate collector is 200～250℃. As a pre-heating device, it is widely used in industrial, mining and petrochemical sectors. The collector has low heat loss and good peak performance. The heat production per unit area can reach 500 W at 220℃. The average annual heat production is high. By absorbing the scattered light, the heat production increases by 30%. The collector has a long life and does not decline in performance, without late maintenance, which is convenient for solar thermal field design. And it is small in size and easy to install and debug. The structure and performance curve of the TVP vacuum flat plate collector are shown in Figure 9.3.

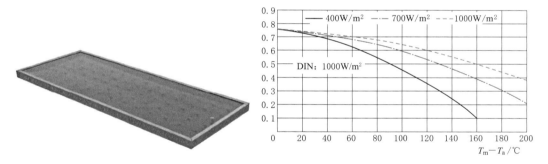

Figure 9.3 TVP vacuum flat plate collector structure and performance curve

> **Specialized Vocabulary**
> - mining 采矿
> - petrochemical 石化的
> - maintenance 维护
> - scattered light 散射光
> - debug 调试

9.1.4 Special-shaped flat plate collector

Chromasun's Micro-Concentrator (MCT) is a new generation of high-performance

Chapter 9 Non-concentrating solar collector

collectors. It uses large flat plate collector technology but achieves a miniaturized package that can be easily integrated into the roof. The efficiency of the collector is higher than that of the ordinary plate collector, and the working medium heating temperature is also higher than that of the ordinary plate collector. The principle of the heat collection is that the sunlight shining on the MCT panel is reflected to the Fresnel lens and concentrates on the absorber for the high-temperature heat collection. The principle is shown in the figure below. The heat sink contains stainless steel tubes that selectively absorb the sunlight. The maximum absorption intensity is 20 times of the ordinary light intensity, and the working medium temperature can reach 204℃. The special-shaped flat collector can be better combined with the building and improve the convenience of using the flat collector (Figure 9.4).

Figure 9.4 The structure of the special-shaped flat plate collector

Specialized Vocabulary

- heat sink   吸热部件
- stainless   不锈钢的
- Fresnel lens   菲涅耳透镜
- miniaturized package   微型包装
- roof   屋顶
- ordinary light   普通光线
- stainless steel   不锈钢

9.1.5 Plastic film-insulated flat plate collector

The new type of flat plate collector applies the highly transparent thin polymer to the flat plate collector. It improves the heat transfer performance of the flat plate collector, making the collector more efficient and reliable. The collector structure is that the plastic film is fixed between the absorber and the anti-reflection glass to prevent the heat loss caused by the convection. Secondly, the back of the absorption device is coated with a film of aluminum powder to replace the mineral cotton used in ordinary collectors. The plastic film between the absorber and the cover plate should have enough space to prevent overheating, which may damage the plastic film.

The Ephex plastic-film flat plate collector from Sweden is designed for household systems. Both the entire collector frame and the absorption surface are made of aluminum. Different from other collectors, the collector's surface is integrated with the whole collector system. The front and back sides of the collector are sealed with polycarbonate. On the back side of the collector, a 4mm thermal reflective light plate is designed, so that its total thickness is only 50mm, the weight is 32 kg, and the total area of the collector plate is 2.03 m^2.

Sunlumo, an Austrian company, has developed all-plastic solar flat-panel collectors, which have obtained the European Solar Keymark certification. The biggest advantage of the new collector lies in its light weight. The weight of the ordinary collector is 15~20 kg/m^2, while that of the Sunlumo all-plastic flat collector is only 8.1 kg/m^2. In addition, the flat-plate collector can withstand high pressure and high temperature as well as the thermal shock for it has the over-heat protection, and its heat loss and collector efficiency can meet the requirements of the testing institutions.

Specialized Vocabulary
- aluminum 铝
- mineral cotton 矿物棉
- household systems 户式系统
- polycarbonate 聚碳酸酯
- transparent 可穿透的
- reliable 可靠的
- overheating 过热
- sealed with 用…密封

9.1.6 Special-shaped flow channel flat plate collector

Savosolar FracTherm plate collector (Figure 9.5) is a typical representative of the special-shaped flow channel plate collector. The collector design is based on mathematical algorithm and adopts a divergent network structure similar to the veins of leaves or human blood vessels. The area of the single collector is 2.15 m^2. The glass of the collector cover plate, the aluminum frame, and the collector backplane and corner are tightly bonded with high-temperature and UV-resistant two-component glue to prevent water vapor from entering the collector. The light transmittance of the collector cover plate is 96.1%, the absorption rate of the heat absorption device is 96%, and the maximum operating temperature of the collector is 134 ℃.

The collector adopts PVD MEMO (Savosolar's MEMO coating process is a self-developed and patented PVD-PECVD coating which takes place in a vacuum chamber and covers the absorbers with three ceramic nanolayers of superior thermo-mechanical and

Chapter 9 Non-concentrating solar collector

Figure 9.5 The structure of Savosolar FracTherm plate collector

optical properties: the coating has an absorbance of 96% and an emission of 5%) as the heat absorption device, having higher efficiency and solar absorption rate, together with the good protective film and exhaust components, which greatly reduce the collector's dust pollution. The collector frame is made of the corrosion-resistant anodized aluminum, with two temperature sensors at the top. The anti-reflection glass cover plate, the aluminum frame, the back absorber and corner end assembly are bonded together with high-temperature anti-UV adhesive to prevent the liquid from passing through the collector.

Specialized Vocabulary

- algorithm 算法
- network structure 网状结构
- veins 脉络，纹理
- anodized 阳极氧化
- divergent 发散的
- transmittance 透光率
- vacuum chamber 真空室
- corrosion-resistant 抗腐蚀
- anodized aluminum 阳极电镀铝
- ceramic nanolayers 陶瓷纳米层
- thermo-mechanical 热力性能
- anti-reflection 抗反射
- adhesive 黏着的

9.1.7 Serpentine flow channel flat plate collector

The main structure characteristic of the serpentine flow channel flat plate collector is that the serpentine coil is used as working fluid channel inside the flat plate collector. As shown in Figure 9.6, the design of the serpentine coil greatly improves the collector efficiency, making it withstand high temperature and have a better heat transfer function. The collector can be installed on flat roofs, sloping roofs or facades. The light transmittance of the collector cover plate is 91%, the absorption rate of the heat absorption device is 95%, and the maximum operating temperature of the collector is about 200 ℃.

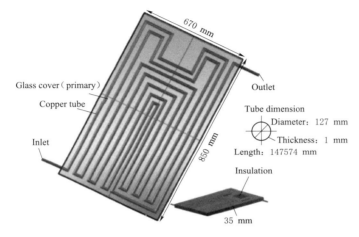

Figure 9.6 The structure of serpentine flow channel flat plate collector

Specialized Vocabulary

- serpentine 蜿蜒, 蛇行
- coil 圈
- sloping 倾斜的
- facade 立面

9.1.8 Flat plate air collector

There are two main types of solar flat air collector: non-permeable air collector and permeable air collector. The air of the non-permeable air collector cannot pass through the suction plate, but flows in the front and back of the suction plate, and exchanges heat with the suction plate heated by solar energy. Compared with the non-permeable air collector, permeable flat air collector adopts the porous heat absorption plate, so the heat transfer is more effective.

The solar wall can generate 600 W of heat per square meter per year, and the sunlight hits the collector, which absorbs solar heat to heat the air, and passes the heated air through tiny pores into the airflow channel. On clear days, the solar flat air collector can raise the air 16~38℃, and finally the solar heating air goes into the building's venti-

Chapter 9 Non-concentrating solar collector

lation system to provide heating for the building. Flat air collectors also help improve indoor air quality and reduce greenhouse gas emissions because they directly heat the fresh air outside.

> **Specialized Vocabulary**
> - permeable 可渗透的
> - non-permeable 不可渗透的
> - suction plate 吸热板
> - porous 多孔的
> - tiny pores 微孔
> - ventilation 通风
> - air quality 空气质量
> - greenhouse gas emission 温室气体排放

9.2 Evacuated tube solar collector

9.2.1 All-glass evacuated tube collector

Vacuum tube collectors are mainly divided into all glass vacuum tube collector and heat tube vacuum tube collector (Figure 9.7). An all-glass evacuated tube solar collector is composed of multiple all-glass vacuum solar collector tubes inserted into the header, because the vacuum tube adopts the vacuum insulation, so the solar thermal energy into the glass tube is not easy to be lost. Therefore, the heat dissipation loss is significantly reduced compared with the flat plate collector, and it still has high thermal efficiency at the operating temperature above 60 ℃. In cold winter, it can still collect heat and achieve a higher thermal efficiency. Due to the advantages of good thermal insulation performance, high low-temperature thermal efficiency and low cost, evacuated tube solar collectors are suitable for use in northern China, and are widely used in household solar water heaters, accounting for about 80% of the domestic market.

Figure 9.7 All-glass evacuated tube solar collector

Since it is made of glass, the all-glass vacuum tube solar collector has a high probability of being damaged when placed outdoors. On the one hand, the broken vacuum tube will cause the dielectric leakage. On the other hand, if one tube is damaged during the operation, the whole system will stop working. Due to the rubber seal between the vacuum tube and the header, the pressure bearing capacity of the system is low. In addition, there are freeze cracks caused by the impact of the hot and cold explosion pipe problems. In order to solve the shortcomings of the all-glass evacuated tube collectors, the improved evacuated tube collectors include U-shaped evacuated tube collectors and heat-pipe evacuated tube collectors. Because there is no liquid flow in the vacuum tube, the U-shaped tube collector has no explosion leakage problem and can operate under pressure. However, the thermal efficiency decreases significantly and the cost increases significantly. Due to the large resistance of the system, the circulatory mediator is easy to overheat and vaporize.

Specialized Vocabulary

- dissipation　耗散
- domestic market　国内市场
- dielectric leakage　介质泄露
- rubber seal　橡胶密封
- shortcoming　缺点
- circulatory mediator　循环介质

9.2.2　Heat pipe evacuated tube collector

Heat pipe evacuated solar collector (referred to as heat pipe evacuated tube) is a kind of solar evacuated collector with metal heat absorbers. It has good heat collection and insulation performance of all-glass evacuated tube, and good thermal conductivity and mechanical strength of the flat plate collector metal. With heat pipe as the core heat transfer element, it has irreplaceable advantages over other types of solar collectors. At present, there are three types of heat pipe evacuated tubes (Figure 9.8). The first one uses hot pressure sealing technology to encapsulate the copper-water heat pipe with flat fins and frost resistance in a single layer evacuated tube. In the second one, the heat pipe with cylindrical fins is encapsulated in an all-glass evacuated tube, through which the hot air is transferred to the fins close to the inner tube surface. In the third way, the heat pipe is encapsulated in an all-glass evacuated tube, and the liquid heat transfer medium with high boiling point is injected between the evacuated tube cavity and the outer wall of the heat pipe.

The heat pipe evacuated tube has a history of more than 20 years in the field of solar thermal energy. Due to its high

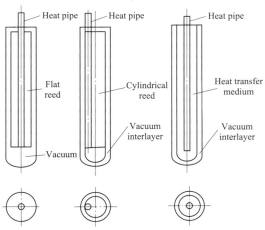

Figure 9.8　Heat pipes of different structures

production technology requirements, complex production process, and copper and main materials of aluminum, the cost is higher than the glass evacuated tube, so its market application is relatively less. At present, with the expanding application of the solar collector and the increasing market demand, the heat pipe evacuated tube has been gradually recognized by the public.

> **Specialized Vocabulary**
> - irreplaceable 不可替代的
> - encapsulated 密封的
> - boiling point 沸点
> - flat fins 平翅片
> - frost resistance 抗寒性
> - mechanical strength 机械强度

Heat pipe evacuated tube is mainly composed of heat pipe (including condensation section and evaporation section), metal heat absorption plate with the selective absorption coating on the surface, glass evacuated tube, degasser, spring support and other accessories, and its basic structure is shown in Figure 9.9.

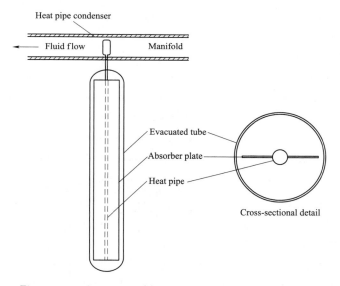

Figure 9.9 Structure of heat pipe evacuated tube collector

The heat pipe used on the evacuated tube is copper gravity heat pipe (also known as thermal siphon, characterized by no liquid suction core in the tube). The heat pipe is filled with special heat pipe working medium inside, and closed to vacuum after exhaust treatment. The heat pipe is divided into two parts: condensation section and evaporation section by the metal cover welded on it. The metal cover and glass tube are sealed to form an evacuated tube after the exhaust treatment. The heat pipe evaporation section and the metal heat absorption plate welded on it are encapsulated in the evacuated tube. The evaporation section of the heat pipe is used to absorb the solar heat collected by the heat absorption plate and transfer it to the condensing section, which then transfers the solar heat to the heated working medium.

The metal heat absorption plate is the core component of the heat pipe evacuated tube, and its performance directly determines the thermal performance of the tube. At present, the heat collecting plate used for the heat pipe evacuated tube in the market is generally oxygen-free copper absorption plate or copper-aluminum composite heat absorption plate, and the surface is formed by magnetron sputtering technology to form the selective absorption coating with high absorption rate and low emissivity. The heat absorption plate and heat pipe are combined or nested together by ultrasonic welding or laser welding to ensure rapid heat conduction.

The sealing of metal cover and glass tube with different expansion coefficient is a key technology of the heat pipe vacuum tube. The glass-metal sealing technology can be roughly divided into two kinds: one is fusion sealing, also known as fire sealing, referring to the use of a thermal expansion coefficient between the metal and glass between the transition material, and the use of flame to melt the glass sealed together; the other is hot pressing sealing, also known as solid sealing, referring to the use of a good plastic metal as the solder, under the condition of heating pressure to seal the metal cap and glass tube together. Because the hot pressing sealing technology has the advantages of low sealing temperature, high sealing speed, low material matching requirements, so at present, in the application of most of the domestic glass-metal sealing technology, the materials used by the solders for hot pressing are plumbum and aluminum.

In order to keep the good vacuum performance of the evacuated collector tube for a long time, both the evapotranspiration degassing agent and non-evapotranspiration degassing agent should be placed in the evacuated collector tube of the heat pipe. The evapotranspiration degassed agent is evaporated on the inner surface of the glass tube after high frequency excitation, like a mirror, and its main function is to improve the initial vacuum degree of the evacuated collector tube. Non-evapotranspiration degassing agent is a long-term degassing agent activated at the room temperature. Its main function is to absorb the residual gas released by various components in the tube and maintain the long-term vacuum of the evacuated collector tube.

Specialized Vocabulary

- condensation 冷凝
- degasser 消气剂

Chapter 9 Non-concentrating solar collector

- spring support 弹簧支架
- accessories 配件
- siphon 虹吸管
- welded 焊接
- magnetron sputtering 磁控溅射
- ultrasonic 超音速的
- fusion 熔化
- solid sealing 固态封接
- plumbum 铅
- evapotranspiration 蒸散
- residual 剩余的

Reading Materials

1. *The dynamic performance of different configurations of solar aided power generation* (SAPG)

2. *Ultrahigh-efficiency solar energy harvesting via a non-concentrating evacuated aerogel flat-plate solar collector*

Reading Materials

Exercises and Discussion

1. What are the structural characteristics of the serpentine flow plate collector?

2. What is the working principle of the flat-plate solar collector?

3. How to reduce the heat loss of the cellular flat plate collector?

4. What can be done to improve the heat collection efficiency of the flat type solar collector?

Chapter 10

Solar thermal energy storage materials

10.1 Overview of solar thermal energy storage materials

Common heat storage methods can be generally divided into three kinds, namely sensible heat storage, phase change heat storage and chemical reaction heat storage. Correspondingly, heat storage materials can also be divided into three kinds, respectively sensible heat storage materials, phase change heat storage materials, chemical reaction heat storage materials.

Sensible heat storage is to store and release heat through the temperature change of heat storage materials. The size of sensible heat storage can be measured by the temperature rise of heat storage materials. The heat storage capacity of sensible heat storage material is reflected in its specific heat capacity. For the same mass of heat storage materials, the greater the specific heat capacity, increase the same temperature to store more heat. Sensible heat storage materials usually have high heat capacity to offset their low energy storage density and low efficiency. Moreover, sensible heat storage has the advantages of simplicity and low cost, and the technology of sensible heat storage is relatively mature at present, which is mainly used in the recovery and utilization of industrial waste heat and the utilization of solar energy. Some sensible heat storage materials commonly used in industry include rock, metal, cement, water and molten salt, etc., which are respectively applied in different occasions and have different temperature ranges.

Phase change heat storage materials are also called latent heat storage materials. Phase change heat storage can absorb or release a large amount of energy under certain conditions. Phase change heat storage has the advantage of high energy storage density, which is much higher than sensible heat storage. In recent years, there are more and more researches on phase change heat storage materials at home and abroad, which has become a hot research direction. Because the performance of phase change materials determines the efficiency of the energy storage device, the efficiency of the heat storage device can vary greatly for different phase change materials. At present, phase change materials used in engineering include paraffin, nitrate, metal and so on.

Chapter 10 Solar thermal energy storage materials

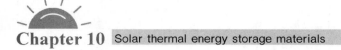

Chemical reaction heat storage is the use of reversible chemical reaction, through the conversion of heat energy and chemical energy to achieve heat storage, with high heat storage density, long heat storage time, can be long distance transmission and other advantages. However, at present, the chemical reaction heat storage is still in the stage of experimental research, and there are still some problems to be solved in the real commercial use.

In a word, heat storage technology is very important for the storage and utilization of solar heat. At present, the sensible heat storage technology has been very mature, but its disadvantage is that the energy storage density is low, which leads to the large volume of the device. In addition, the efficiency of the sensible heat storage technology is also low, so the sensible heat storage technology has certain limitations. Chemical reaction heat storage is still in the stage of experimental research, but the research progress in this area is relatively fast. At present, phase change energy storage technology has attracted a lot of research due to its advantages and strong development momentum. However, traditional phase change materials have many problems in practical application, such as low thermal conductivity. In addition, the cost of phase change energy storage is also relatively high, which needs to be solved before large-scale commercial use. However, in recent years, with the emergence of new phase change materials such as nano-composite phase change heat storage materials, the above problems are expected to be solved.

Specialized Vocabulary

- solar thermal energy storage material 太阳能储热材料
- sensible heat storage 显热储热
- phase change storage 相变储热
- chemical reaction storage 化学反应储热
- specific heat capacity 比热容
- paraffin 石蜡
- nano-composite phase change materials 纳米复合相变储热材料
- phase change 相变

10.2 Sensible heat storage material

10.2.1 Liquid sensible heat storage

At present, the widely used liquid heat storage media include all kinds of molten salt, mineral oil, liquid metal and water.

Water is one of the best and most widely used liquid heat storage media. Water has the advantages of high specific heat capacity, low cost and non-toxicity. As a result, water is now widely used for heating in urban areas. In addition, water as a heat storage medium is also used in solar collectors in many rural areas of China. However, water also has some disadvantages, such as its high vapor pressure, as well as the problems of sealing and corrosion.

Molten salt has better heat storage and heat transfer performance, and its working temperature matches that of steam turbine with high temperature and high pressure. It is liquid under atmospheric pressure, not easy to burn, no toxicity, and lower cost, and is more suitable for high temperature the physical properties of some molten salts are shown in Table 10.1 power generation. The widely used molten salts are binary and ternary mixed molten salt. The physical properties of some molten salts are shown in Table 10.1. The disadvantages of molten salt are high temperature corrosion and low temperature solidification and other problems. The relevant pipe valve materials must be resistant to high temperature and corrosion, and the relevant equipment must be insulated and preheated. Generally speaking, the molten salt heat storage material needs to be improved in its fluidity, high temperature stability, temperature range and other aspects.

Table 10.1 The melting point and boiling point of the commonly used molten salt

Composition	Melting point	Boiling point/℃
$NaNO_3$	310	380 (decomposition)
$NaNO_2$	270	320 (decomposition)
KNO_3	337	537
Na_2NO_3	851	1600
60% wt $NaNO_3$ ~ 40% wt KNO_3	220	590
NaCl	801	1465
KCl	770	1420

Mineral oil is also an ideal liquid heat storage medium. Mineral oils have a lower vapor pressure than water and are capable of operating as a liquid at temperatures up to 400℃. Also, unlike molten salt, mineral oil does not freeze in pipe at night, but it is more expensive than molten salt.

Pure metals and alloys also have the potential as sensible heat storage media. They have a high thermal conductivity and maximum operating temperature. Their vapor pressure is minimal. But they have drawbacks like the high cost. Also, they may require oxygen and oxide-free environment in order to reduce the corrosion.

10.2.2 Solid sensible heat storage

Solid sensible heat storage materials have low cost and are easily available everywhere. They have no vapor pressure issue for the operating pressure is close to the ambient pressure without the need of the pressure containing vessels or leak issues. Since they cannot be circulated easily, they can only do passive heat storage and need a fluid, usually air (HTF) for transporting heat into and out of the loosely packed solid materials in a storage tank. To increase the heat transfer efficiency, there is a direct contact between the flowing air and solid heat storage medium during the charging and discharging process. One of the issues with sensible storage is that during the discharging process the temperature of the storage medium decreases, so the HTF temperature also decreases with time. The low temperature applications like space heating and industrial waste heat recovery may use some of the solid sensible heat storage materials. The cylindrical honeycomb ceramic heat accumulator is shown in Figure 10.2.

Figure 10.1 Molten salt in a solid state that has not reached the melting point

Figure 10.2 Cylindrical honeycomb ceramic heat accumulator

Specialized Vocabulary

- molten salt 熔盐
- non-toxicity 无毒性
- corrosion 腐蚀
- atmospheric pressure 大气压
- solidification 凝固
- mineral oil 矿物油
- drawback 缺点
- Melting point 熔点
- loosely 宽松地
- cylindrical honeycomb ceramic 圆柱形蜂窝陶瓷

10.3 Phase change materials

Phase change materials (PCMs) are substances that release or absorb a lot of heat through phase change to achieve heating or cooling. The latent heat of phase change material is much larger than its sensible heat. For example, for water, 1g of ice needs to absorb 333.55J of heat to melt into water, while 1g of water needs to absorb 4.18 J of heat to raise its temperature by 1K. Therefore, under certain conditions, water is a very ideal phase change material.

Phase change materials are mainly divided into organic phase change materials and inorganic phase change materials. Among them, only solid – liquid phase variable materials are more realistic, and are most used at present. For the liquid to gas phase transformation, due to the large volume of gas, high pressure storage is required, so the requirements for the storage device are relatively high, such as the requirements for the sealing of the device is relatively high. Solid – solid phase transitions are less useful because the reaction is very slow and less heat is released during the transition. The classes of existing PCMs are shown in Figure 10.3.

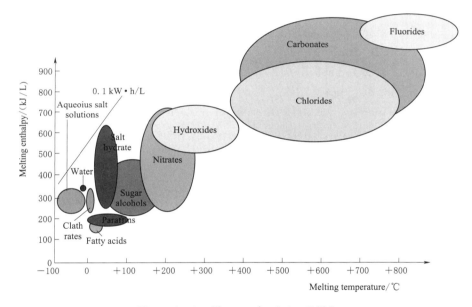

Figure 10.3 Classes of existing PCMs

10.3.1 Organic phase change material

Organic phase change energy storage materials usually have better thermal properties and chemical energy than the inorganic phase change energy storage materials, such as stable, non – toxic and non – corrosive, so the organic phase change energy storage materials are more likely to be selected by researchers. Organic PCMS mainly include paraffin (Figure 10.4), fatty acids and organic compounds such as alco-

Figure 10.4 Paraffin

hol. Organic PCMS are often cheaper and easier to get. Paraffin is the most representative material in organic PCMS and is widely used in engineering. The phase transition temperature of paraffin is 0~80℃, and latent heat of phase change is as high as 150~250 J/g, paraffin, which is widely valued in the study of phase change energy storage materials because of its large temperature range, wide range of temperature range and low price.

Paraffin or chain alkane is a combination of straight chain alkane, which can be used in a general type. The molecular chain of paraffin can release a lot of potential heat during phase transition, and the melting point of paraffin wax increases with the increase of the chain. The research shows that the paraffin is absorbed, stored and released in a large amount of heat in the heat of the solid and liquid, which is mainly used in the energy storage system of the building energy saving and solar storage. However, paraffin also has low thermal conductivity and small density problems.

The alcohols, which are used as phase change materials, include two alcohol and polyols (sugar alcohol), which are used as phase change materials to be used as a polymer polyethylene glycol (PEG), such as PEG 400, which has good chemistry and thermal stability, non-combustible, non-toxic, non-corrosive, and low price. The melting point and melting of PEG increase with the increase of molecular weight. Polyethylene glycol is widely used in solar, building, daily necessities and aerospace.

10.3.2 Inorganic phase change material

The inorganic type of heat transfer materials are mainly crystalline water and salt, molten salt, metal or alloy. The application of crystalline water and salt is extensive, including brine, sulfate, phosphate and other alkali or alkali halide. Due to the large thermal conductivity, the transition of phase change is high, the price is low, and the industry is used heavily. However, inorganic salt is easily separated from the cold phenomenon, reducing the sensitivity and accuracy of the phase change materials, although there are more applications in the industry, but these shortcomings can hinder the application and promotion of inorganic salt.

Metal or alloy phase change materials are also an important part of inorganic class change materials. This kind of material tend to have high thermal conductivity, and they can be reused many times with little performance loss. They are also more reliable. They have the highest heat of phase transition per unit volume or per unit mass. As a result, they have a high energy storage capacity. During the phase transition, the change in volume is negligible. Their vapor pressure is also negligible. When considering the volume of the system, they can replace salts as the heat storage medium because of their high heat storage density.

10.3.3 Composite phase change materials

Different kinds of phase change thermal storage materials have different characteristics and limitations. Two or more phase change materials can be combined together by a certain composite method to obtain the composite phase change materials with excellent performance. Composite phase change materials (one of them is shown as Figure 10.5) can not only overcome the shortcomings of single inorganic or organic phase change heat storage materials but also improve the application effect of phase change materials and expand their application range. At present, the composite methods of phase change heat storage materials mainly include microcapsule phase change materials, nanocomposite phase change materials and shaped phase change materials.

Nano-composite phase change material is a kind of composite material by adding a certain amount of nano-particles into the phase change material. With the addition of nanoparticles, the thermal conductivity of the phase change material is enhanced, therefore, the heat storage and heat transfer efficiency of the system is also enhanced. Nanoparticles include carbon-based nanoparticles and metal nanoparticles, both of which have high thermal conductivity.

The phase change microcapsule is a new type of energy-saving and environmentally friendly energy storage material that is composed of a polymer with a polymer to form a micro (nano) scale of the nuclear shell. The capsule technology not only avoids the direct contact with the outside world, but also solves the problem of the leakage of phase change materials and phase separation. This technique can prevent the permeability of liquid material in phase transition process, improve the efficiency of phase change materials and expand the application of phase change materials. Different types of graphite composites can be obtained by treating graphite in different ways. There are graphite flakes, expanded natural graphite, expanded graphite powder (Figure 10.6).

Figure 10.5　A high-density polyethylene

Figure 10.6　High temperature expanded graphite powder

Specialized Vocabulary

- latent heat　潜热
- organic　有机的
- inorganic　无机的

Chapter 10 Solar thermal energy storage materials

- fatty acids   脂肪酸
- phase transition   相变
- microcapsule   微胶囊
- alcohols   醇类
- graphite   石墨
- polyethylene   聚乙烯
- graphite flakes   石墨片
- expanded natural graphite   膨胀天然石墨
- expanded graphite powder   膨胀石墨粉
- polymer   聚合物
- polyethylene glycol   聚乙二醇
- extensive   大量的
- non-combustible   不易燃的

10.4 Heat storage by chemical reaction

Chemical heat storage uses reversible chemical reactions to store and release heat. The heat storage density of chemical heat storage is very high, much higher than that of sensible heat storage and phase change heat storage, which is a very big advantage of chemical heat storage. Chemical heat storage can not only store thermal energy for a long time with almost no heat loss, but also realize the combination of cold and hot storage, which is beneficial to the long-term storage of energy. Moreover, for chemical heat storage, there are more heat storage materials or reversible chemical reactions available, and the applicable temperature range is wide. Chemical reaction heat storage has a high application prospect in solar energy heat storage and industrial waste heat recovery and utilization. In addition, chemical heat pumps are expected to replace traditional heat pumps in the future, and the efficiency of chemical heat pumps may be higher. Because many chemical substances have certain dangers, special devices need to be used to isolate them, so the system is complex, large, high investment, and the overall efficiency is still low. In addition, the chemical reaction process is complex, the kinetic characteristics of some reactions are not completely clear, and some reactions need catalysts, which have certain safety requirements, and are still in the stage of small-scale research and trial. There is a technology that uses a chemical reaction to produce hydrogen with an efficiency of up to 30%. It uses solar energy high temperature thermal chemical reaction and metal oxide as a catalyst to prepare hydrogen, chemical reaction process only hydrogen and oxygen, no other substances, this is a very environmentally friendly hydrogen production technology, which has great application prospects.

10.4 Heat storage by chemical reaction

Chemical heat storage is a multidisciplinary integrated technology, if the above problems can be solved well, its application prospect will be very broad.

Specialized Vocabulary

- chemical reaction 化学反应
- reversible chemical reaction 可逆化学反应
- long-term storage 长时间储存
- catalyst 催化剂
- trial 试验
- solar energy high temperature thermal chemical reaction 太阳能高温热化学反应
- hydrogen 氢气
- oxygen 氧气
- environmentally friendly 环保的
- multidisciplinary 多学科

Reading Materials

1. *Review on micro/nano phase change materials for solar thermal applications*
2. *Thermal energy storage and solar energy utilization enabled by novel composite sodium acetate trihydrate/sodium dihydrogen phosphate dihydrate phase change materials*

Reading Materials

Exercises and Discussion

1. What kinds of solar energy storage materials are there?
2. Which material retains the same temperature during the heat storage process?
3. What kinds of graphite composite materials are there?
4. What are the advantages and disadvantages of solid heat storage materials?

Chapter 11
Practical application of solar thermal technology

At present, the application of solar thermal energy is the most commercialized and widely applied scheme in the field of the renewable energy. According to the temperature range of solar energy collection, the thermal application of the solar energy can be divided into: low temperature application of solar energy with heat collection temperature less than 200℃, medium temperature application with that between 200～800℃ and high temperature application with that greater than 800 ℃. Depending on the temperature range, the solar energy can be used for different purposes, and the higher the temperature is, the better the quality of the energy will be. Solar low temperature applications are mainly used in solar water heaters, solar agricultural drying, seawater desalination, solar houses and solar refrigeration systems. Medium temperature applications mainly include solar ovens, solar thermal power generation, and industrial preheating. High temperature applications can be used in high temperature solar furnaces, solar thermal chemistry and other applications. At present, the application of solar thermal energy in various fields has shown its development potential and has a good application prospect.

Specialized Vocabulary

- commercialized 商业化的
- renewable 可再生的
- oven 烤箱
- desalination (海水) 淡化
- solar agricultural drying 太阳能农产品烘干
- solar furnace 太阳炉
- various 各种各样的

11.1 Application of low temperature solar energy

Solar water heater is the largest application form of the solar heat in the industry. The original solar water heater is a black metal drum. On a calm summer day, a single metal container can produce only 38.9 ℃ of hot water in a day. In 1909, Bailey designed a solar hot water system that separated the heat collection and storage parts to form a flat plate collector and a heat storage tank. Solar water heaters can be divided into three types according to the flow mode: circulation, direct flow and integral. Natural and forced circulation are the two main types of circulation in terms of the circulation power. The direct flow is an open thermal siphon system composed of a flat plate collector, a heat storage tank, a supply tank and connecting pipes. The integral type is the collector and the heat storage tank are integrated into one, which is the most original solar water heating device. The schematic diagram shows three kinds of solar collector structure in Figure 11.1~Figure 11.3.

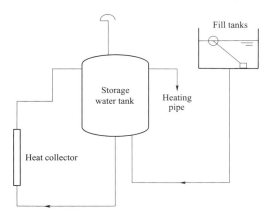

Figure 11.1　Solar natural circulation hot water system

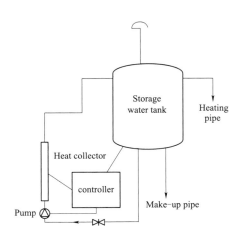

Figure 11.2　Solar forced circulation hot water system

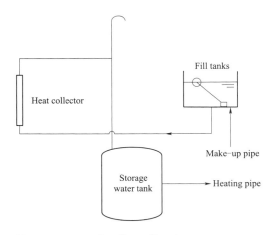

Figure 11.3　solar direct flow hot water system

Chapter 11 Practical application of solar thermal technology

> **Specialized Vocabulary**
> - container 容器
> - integral 整体
> - schematic 示意图
> - solar direct flow hot water system 太阳能直流式热水系统
> - solar forced circulation hot water system 太阳能强制循环热水系统
> - solar natural circulation hot water system 太阳能自然循环热水系统

11.2 Application of medium temperature solar energy

The application of the solar energy in the high temperature is very extensive, in daily life, or in industrial and agricultural production, such as solar oven, solar thermal power generation, seawater desalination and food processing.

11.2.1 Solar oven

A solar oven is a device that collects the solar energy and uses it for cooking or boiling water. The key component of the solar oven is the concentrator, which is not only the choice of mirror material, but also the design of the geometry. The most common mirror is the silver–plated or aluminized glass mirror, and there is also aluminum polished mirror and polyester film aluminized materials (Figure 11.4).

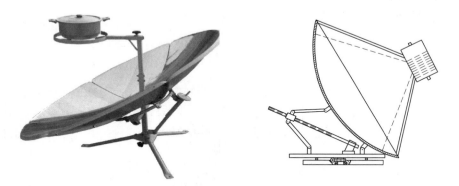

Figure 11.4 Physical drawing and structure drawing of the concentrating solar oven

Solar oven can be divided into three kinds according to the way of concentrating light: box–type solar oven, plate solar oven and concentrating solar oven. Box–type solar oven is developed based on the principle that black objects absorb solar radiation better. It is in the shape of a box, sealed and insulated, with only one side left to receive the sun's rays. Its structure is simple but its operating temperature is low. The flat plate solar oven is a combination of flat plate collector and box solar oven. Concentrating solar ovens use concentrating mirrors to focus a large area of sunlight onto the bottom of a pot for heating.

The advantage of solar ovens lies in their high economic efficiency. It can cook food and heat water without using coal or natural gas. Solar stoves basically do not have any pollution to the environment, green environmental protection. But its disadvantage is restricted by the site, weather and other natural conditions. Not available in rainy weather or indoors.

Specialized Vocabulary

- polyester 聚酯纤维
- aluminized 镀铝的
- box-type solar oven 箱式太阳能灶
- plate solar oven 平板太阳能灶
- concentrating solar oven 聚光太阳能灶

11.2.2 Sea water desalination

Solar seawater desalination is mainly divided into the use of solar thermal energy and photovoltaic electricity to desalination. There are two methods for the seawater desalination phase transformation using the solar energy: direct and indirect methods. The direct method uses the heat energy absorbed by the solar energy collection system to distill seawater directly, while the indirect method separates the heat collection and distillation process and uses the working medium for heat transfer to complete the seawater desalination. At present, the conventional solar desalination system has the following problems: the latent heat of steam condensation generated in the distillation process cannot be effectively utilized; the large amount of circulating seawater and the large total heat capacity weaken the driving force of the evaporation. The distillation system is mainly dominated by the natural convection and the heat transfer efficiency is low. The solar heat collection area is high, the initial investment of the system is too large, and the cost of water production is too high. The low density and instability of the solar energy affect the stable operation of the system. In order to improve the solar desalination system, many new desalination technologies have been developed, some of which are briefly introduced below.

Specialized Vocabulary

- distillation 蒸馏
- Solar seawater desalination 太阳能海水淡化
- latent heat 潜热
- driving force 驱动力
- natural convection 自然对流
- initial investment 初投资

Chapter 11 Practical application of solar thermal technology

11.2.2.1 Solar humidification – Dehumidification seawater desalination

Solar humidification and dehumidification sea water desalination is based on solar energy as the heat source, in the process of air humidification and dehumidification to obtain fresh water. The whole system can be divided into four processes: solar heat collection process, seawater heating process, fresh water precipitation process and air circulation process. First, feed seawater enters the solar heat collection system through the condenser to absorb the solar energy and evaporate after heating. The vaporized seawater is then sent into the humidifier to spray the humidifying air, and the concentrated brine is discharged from the back end. During humidification and dehumidification, air circulates in the closed chamber. In the condenser, the air condenses and dehumidifies to produce fresh water which is collected into the fresh water tank; The air then enters the humidifier to humidify and heat up before returning to the air conditioner and entering the cycle. Humidification and dehumidification seawater desalination system has simple operation, atmospheric pressure operation, stable water production and so on. Using solar energy as power source can not only save the operating cost of the system, but also reduce the pollution to the outside world.

> **Specialized Vocabulary**
> - humidification 加湿
> - dehumidification 除湿
> - chamber 腔体
> - humidifier 加湿器
> - concentrated brine 浓缩盐水

11.2.2.2 Seawater desalination by solar membrane distillation

Membrane distillation is a membrane separation process using latent heat of evaporation to realize phase transformation. The mass transfer force of evaporating components is the vapor pressure difference on both sides of the hydrophobic distillation film. The operation process is that the vacuum system is used to continuously vacuum the membrane side to maintain the steam partial pressure difference between the hot seawater side and the membrane side, and then the steam pumped out by the membrane side is condensed to desalinate water. The technology is organically combined with solar energy to form a solar membrane distillation system.

At present, solar membrane distillation system is mainly divided into two forms. One is a direct heating system, which uses solar energy to heat seawater directly and send it into a distillation membrane to separate steam from liquid and condense the steam into fresh water. The other uses the intermediate heat transfer medium, first heating the intermediate heat transfer medium with solar energy, and then the heat transfer medium and

11.2 Application of medium temperature solar energy

water seawater heat desalination.

A solar membrane distillation system usually consists of a solar collector, membrane assembly, condenser, and other auxiliary equipment, as shown in Figure 11.5. The advantages of the system are: low operating pressure, low operating temperature, easy to operate, low operating risk; High purity of produced water, green environmental protection.

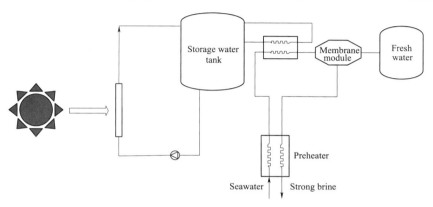

Figure 11.5 Solar membrane distillation seawater desalination system

Specialized Vocabulary

- hydrophobic distillation film 疏水蒸馏膜
- membrane 膜
- intermediate heat transfer medium 中间换热介质
- assembly 组件
- purity 纯度

Reading Materials

1. *Practical design and construction of solar ponds*
2. *Review on sensible thermal energy storage for industrial solar applications and sustainability aspects*

Reading Materials

Exercises and Discussion

1. What are the examples of solar low temperature applications?
2. What is the difference between solar humidification-desiccant desalination and solar membrane distillation desalination?
3. Which of the three forms of solar ovens is the most efficient in collecting heat?

Appendix

List of specialized words

Photovoltaic

a rear layer	底板
absorb $v.$	吸收
absorption coefficient	吸收系数
AC load	交流负载
active area	有效面积
Aluminium doped zinc oxide (AZO)	掺铝氧化锌
amorphous silicon	非晶硅
anatase $adj.$	锐钛矿（型）
anode/photoanode film	阳极/光阳极薄膜
antibonding orbital	反键轨道
antidumping, countervailing	反倾销、反补贴（双反）
antireflection coating (ARC)	抗反射层
antireflection film	减反射膜
Atmospheric Pressure Chemical Vapor Deposition (APCVD)	常压化学气相沉积
atomic layer vapor deposition	原子层气相沉积
back electrode	背电极
band gap tailoring	带隙调整
band gap	带隙、能隙、能带间隙
base region	基区、基极区
battery (powered) car	电瓶车
battery $n.$	蓄电池
biogas $n.$	生物气，沼气
buffer layer	缓冲层
building attached photovoltaics (BAPV)	光伏与建筑结合/一体化
building integrated photovoltaics (BIPV)	光伏建筑一体化
built–in battery	内置蓄电池
built–in potential	内建电势，内建电场，内建电位
built–in	内置

busbars *n.*	母线
cadmium telluride (CdTe)	碲化镉
capacity *n.*	容量
carbon negative	碳负极
carrier density	载流子密度
carrier *n.*	载流子
cast multicrystalline silicon	铸造多晶硅
cast *v.*	浇铸
cell packaging	电池封装，电池包装
centralized electric generation system	集中式发电系统
charge controller	充放电控制器
charge recombination	电荷复合
chemical bath deposition	化学浴沉积法
chemical vapor deposition (CVD)	化学气相沉积
close – space sublimation (CSS)	近空间升华法
coal *n.*	煤
combined heat and power	热电联供
concentration ratio	聚光比，聚光比例
concentrator solar cell	聚光太阳电池
conduction – band	导带，传导带，导电带
conductive substrate	导电基底
conjugated system	共轭体系
convert into power	转化为电力
converting solar energy into electricity	太阳能转换成电能
cost – effective	划算的，成本效益好的，合算的
counter electrode	对电极
covalent bond	共价键
current density	电流密度
cut carbon dioxide emissions	二氧化碳减排
czochralski silicon (Cz – Si)	直拉单晶硅
czochralski technique (CZ)	直拉法
dangling bonds	悬挂键，悬空键
daylighting import system	采光导入系统
delocalized bonding	离域键，非定域键
dense layer	致密层
depletion heterojunction quantum dot solar cells	耗尽异质结量子点太阳电池
depletion region	耗尽区层，势垒区，阻挡层
device operation	器件运行，器件工作，设备操作

diesel generator	柴油发电机
diffusion length	扩散长度
diffusion $v.$	扩散
direct current (DC)	直流
direct current	直流、直流电
direct-coupled system	直接耦合系统
distribution panel	配电盘
dominant energy	主导能源
doping $v.$	(在半导体材料中)掺杂
dual-junction	双结
dye-sensitized solar cells (DSSC)	染料敏化太阳电池
edge insulation processing	边缘绝缘处理
efficient crystalline silicon solar cells	高效晶体硅太阳电池
efficient monomer batteries	高效单体电池
efficient solar modules	高效太阳电池组件
electric field	电场
electrolyte $n.$	电解液,电解质
electron-hole pairs	电子-空穴对
(electron) transfer efficiency	转换效率
electron transport layer	电子传输层
emergency charger	应急充电器
encapsulant $n.$	密封材料,密封剂
encapsulation material	封装材料
energy consumption	能耗
energy intensive manufacturing process	能源密集型制造过程
energy mix	能源组成
environmental protection	环境保护
epitaxial growth	外延生长
epitaxial method	外延法
epoxy resin encapsulation	环氧树脂封装
ethyl vinyl acetate (EVA)	乙烯基醋酸乙酯
ethylene tetrafluoroethylene (ETFE)	聚氟乙烯
evaporation sputter	蒸发溅射
exciton $n.$	激子
fabrication processes	制备工艺,合成过程
feedstock $n.$	原料
fill factor	填充因子
flat solar cells	平板太阳电池

flexible perovskite solar cells	柔性钙钛矿太阳电池
Float Zone method (FZ)	区熔法
Fluorine doped tin oxide (FTO)	掺氟氧化锡
forbidden band	禁带
forward–current	正向电流
fossil fuel–fired generation	化石燃料发电
fossil fuels	化石燃料
foundation materials	基础材料
Fresnel lenses	菲涅尔透镜
generate electricity	发电
generator $n.$	发电机，发生器，产生器
geothermal $n.$	地热
germanium $n.$	锗
global positioning system (GPS)	全球定位系统
grain boundaries	晶界
grain size	晶粒尺寸
grid line	栅线
grid $n.$	输电网
grid–connected generation	并网发电
grid–connected photovoltaic power systems	并网光伏发电系统
grid–connected	并网（发电）
ground application	地面应用
heat flux	热通量、热流
heterojunction devices	异质结器件
heterojunction $n.$	异质结
heterostructure $n.$	异质结构，异质结
high brightness	高亮度
highest occupied molecular orbital (HOMO)	最高占有分子轨道
high–rate semiconductor deposition	高效/高速半导体沉积
hole transport layer	空穴传输层
holographic concentrators	全息聚光，全息聚光器
homogenous layer	均质层，均匀层
homogenous $adj.$	同质的
homojunction $n.$	同质结
hybrid system	混合系统
hydrocarbons $n.$	碳氢化合物，烃类
indirect band gap	间接带隙
indium gallium phosphide	铟镓磷

indium tin oxide (ITO)	铟锡氧化物
infrared radiation	红外线照射
ingot *n*.	铸块、锭
inorganic compounds thin-film solar cells	无机化合物薄膜太阳电池
inorganic-organic lead halide perovskite	有机-无机卤化铅钙钛矿
installed capacity	装机量，安装量
intermetallic *n*.	金属间化合物
International Renewable Energy Agency (IRENA)	国际可再生能源署
intrinsic concentration	本征浓度
inverse electron recombination	电子反向复合过程
inverter *n*.	逆变器
Inverter/Power conditioner	功率调节器
laminated packaging	层压封装
lattice mismatch	晶格失配
launched into market	投放市场，投入市场
lead-acid battery	铅酸蓄电池
lenses *n*.	透镜
light emitting efficiency	发光效率，出光效率
light sensors	光感应器
lighting circuit	照明电路
lighting lamps	照明灯管
liquid phase deposition	液相沉积法
load resistance	负载电阻
long-range order	长程有序
low power-consumption	低功耗
lowest unoccupied molecular orbital (LUMO)	最低空分子轨道
luminescent *adj*.	发光，发光的，冷光的，场致发光
maximum power output	最大输出功率
maximum power point trackers (MPPTs)	最大功率点跟踪
mesoporous perovskite solar cells	介孔钙钛矿太阳电池
mesoporous *n*.	介孔、中孔
metal electrode	金属电极
mid-latitude area	中纬度地区
mirrors *n*.	反射镜、反光镜
modules *n*.	组件，模组，模块
molding methods	成型方法
molten silicon	熔融硅
monocrystalline silicon boron back surface field	单晶硅硼背场

monocrystalline silicon	单晶硅
multifunctional solar emergency charger	多功能太阳能应急充电器
multi-junction cells	多结电池
multi-junction	多结
multimeter *n.*	万用表
MW, megawatt	兆瓦，百万瓦特
nanocrystalline *n.*	纳米晶（体）
National Renewable Energy Laboratory (NREL)	国家可再生能源实验室
natural gas	天然气
negative charge	负电荷
northern hemisphere	北半球
ohmic contact	欧姆接触
oil, petroleum	油，石油
one-axis *n.*	单轴
open-circuit voltage	开路电压
optical absorption coefficients	光吸收系数
optical lens	光学透镜
optical sensing	光学传感
optoelectronic properties	光电性能
organic solar cells/organic photovoltaics (OPV)	有机太阳电池
organometal halide perovskites	有机金属卤化物钙钛矿
organo-metallic chemical vapor deposition (OMCVD)	有机金属化学气相沉积
over discharge	过放
overcharge *v.*	过充
perovskite absorption layer	钙钛矿吸收层
perovskite solar cells (PSC)	钙钛矿太阳电池
photoelectric catalysis	光电催化
photolysis *n.*	光解作用
photosensitization *n.*	光敏作用
photosensitizers *n.*	光敏剂
photosynthesis *n.*	光合作用
photovoltaic concentrators	光伏聚光器
photovoltaic curtain wall	光伏幕墙
photovoltaic effects	光伏效应
photovoltaic power station	光伏电站
photovoltaic subsidy policies	光伏补贴政策
Physical Vapor Deposition (PVD)	物理气相沉积

planar donor – acceptor heterojunction	平面供体-受体异质结
planar perovskite solar cells	平面/板钙钛矿太阳电池
Plasma Enhanced Chemical Vapor Deposition (PECVD)	等离子体增强化学气相沉积
plasma etch	等离子体刻蚀
plateau ['plætəʊ] area	高原地区
p – n junction	p – n 结
polarity *n.*	极性
poles *n.*	灯杆
polycrystalline silicon	多晶硅
polymer reflectors	聚合物反射镜
polyvinyl fluoride (PVF)	聚氟乙烯
polyvinyl – butiral (PVB)	聚乙烯醇缩丁醛
positive charge	正电荷
power conditioning units	功率调节单元，功率调节器
power output	输出功率
processing conditions	加工条件、制造条件
PV corridor	光伏走廊
PV power generation	光伏发电
PV system	光伏系统
quantum dot polymer hybrid solar cells	量子点聚合物杂化太阳电池
quantum dot schottky solar cells	量子点肖特基太阳电池
quantum dot sensitized solar cells	量子点敏化太阳电池
quantum dots solar cells (QDSC)	量子点太阳电池
radiation resistance	耐辐射性，辐射抗性
Radio Frequency Sputtering	射频溅射
recombination loss	复合损失
rectifier *n.*	整流器
redox *v.*	氧化还原，氧化还原剂，氧化还原反应
reflective *v.*	反射，反光
regional distribution	地域分布，区域分布
residential grid – connected photovoltaic power systems	住宅光伏并网发电系统
reverse charging	反充
ribbon silicon	带硅
rigid frame	钢架，刚性构架
rovers *n.*	火星车
sail *n.*	帆

satellite applications	卫星应用
selenization *n*.	硒化法
short – circuit current	短路电流
shunting path	分流路径，分流电路
single – junction devices	单结器件
single – junction	单结
sinter *v*.	烧结
small – scale power system	小规模电源系统
soda lime glass	钠钙玻璃
solar (power generation) projects	太阳能发电项目
solar array/PV array	太阳电池阵列/光伏阵列
solar cell module	太阳电池组件
solar desalination [diːsælɪ'neɪʃən]	太阳能海水淡化
solar devices	太阳能装置
solar drying/solar energy drying	太阳能干燥
solar electric car	太阳能电动车
solar energy distribution	太阳能分布
solar energy radiation quantity	太阳辐射量
solar energy resource abundance	太阳能资源丰度
solar energy resource(s)	太阳能资源
solar energy utilization patterns	太阳能利用方式
solar flux	太阳辐射通量
solar garden light	太阳能草坪灯
solar light	太阳能灯
solar modules	太阳电池组件
solar panel	太阳电池板
solar photocatalysis	太阳能光催化
solar power	太阳能发电
solar radiation	太阳辐射
solar street light	太阳能路灯
solar thermal power generation	太阳能热发电
solar thermal	光热，太阳能热，太阳热
solar traffic warning lights	太阳能交通警示灯
solar water heater	太阳能热水器
solar yacht	太阳能游艇
solar – grade silicon	太阳级硅
solar – power conversion efficiency	光电转换效率，太阳能转换效率
space application	空间应用

space power	空间电源
speckled crystal reflective appearance	斑点晶体发光的外观
sphalerite crystal structure	闪锌矿晶体结构
spin – coating	旋涂，旋涂仪
sputtering $v.$	溅射
stand – alone PV system	独立光伏发电系统
State Grid	国家电网有限公司
sunlight flux	阳光通量
sun – tracking system	太阳跟踪系统
superstrate $n.$	覆盖物
sustainable light	可持续发光
switch on/off	开/关
technical barriers	技术瓶颈，技术壁垒
tedlar $n.$	泰德拉（宇宙飞船用以保留热量的绝缘材料），聚氟乙烯
terrestrial power generation	地面发电
texture etch	织构刻蚀，表面刻蚀
texturing $adj.$	织构化
the junction area	结区
the metal grid contact	金属栅接触
the parietal area	顶叶区
the top region	顶区
the total capacity	总发电量
thermoplastic polyurethane (TPU)	热塑性聚氨酯
threshold $n.$	阈值
total energy consumption	总能耗，能源消费总量
total energy structure	总能源结构
total solar sunshine hours	太阳日照总时数
triple – junction	三结
two – axis tracking systems	双轴跟踪系统
ultraviolet irradiation	紫外线照射
utility grid	公用电网
valence band	价带
volt – ampere characteristic	伏安特性
wafer slicing	晶圆切片，硅片切割

Solar Thermal

absorption refrigeration	吸收式制冷
accessories *n*.	配件
adaptability *n*.	适应性
additional sunspace type	附加阳光房式
additional sunspace type	附加阳光间式
adhesive *adj*.	黏着的
adobe *n*.	黏土
adsorption *n*.	吸附
air convection	空气对流
air quality	空气质量
alcohols *n*.	醇类
algorithm *n*.	算法
aluminized *adj*.	镀铝的
aluminum *n*.	铝
ammonia *n*.	氨
ammonia – water *n*.	氨水
annual cost	费用年值
anodized aluminum	阳极电镀铝
anodized *v*.	阳极氧化
anti – reflection *n*.	抗反射
aperture receiver	孔径接收器
aquifer heat storage	含水层蓄热
assembly *n*.	组件
atmospheric pressure	大气压
auxiliary energy device	辅助能源装置
biomass boiler	生物质锅炉
boiling point	沸点
box – type solar oven	箱式太阳能灶
brick *n*.	砖
buffer *n*.	缓冲区
buried pipe heat storage	地埋管蓄热
catalyst *n*.	催化剂
cellular flat plate collector	蜂窝式平板集热器
ceramic nanolayers	陶瓷纳米层
chamber *n*.	腔体

chemical heat storage	化学蓄热
chemical reaction storage	化学反应储热
chemical reaction	化学反应
Chinese Academy of Sciences	中国科学院
circulatory mediator	循环介质
cluster $n.$	群、束
coil $n.$	圈
combination type	组合式
combination type	组合式
commercialized $adj.$	商业化的
compact structure	紧凑结构
concentrated brine	浓缩盐水
concentrated solar power generation technology	聚光太阳能发电技术
concentrated solar power plant	聚光型太阳能光热发电
concentrating solar oven	聚光太阳能灶
concentrating solar power	聚光太阳能电站
concentration ratio	聚光比
concentrator $n.$	集热器
concrete $n.$	混凝土
condensation $n.$	冷凝
container $n.$	容器
convection $n.$	对流
conventional energy	传统能源
conversion efficiency	转换效率
corrosion $n.$	腐蚀
corrosion – resistant	抗腐蚀
cylindrical honeycomb ceramic	圆柱形蜂窝陶瓷
debug $v.$	调试
degasser $n.$	消气剂
dehumidification $n.$	除湿
demonstration project	示范工程
desalination $n.$	(海水)淡化
desiccant $n.$	干燥剂
dielectric leakage	介质泄露
direct – gain type	直接受益式
discretization $n.$	离散化
disk solar thermal power system	碟式太阳能热发电系统
dissipation $n.$	耗散

distillation *n.*	蒸馏
divergent *n.*	发散的
domestic market	国内市场
drawback *n.*	缺点
driving force	驱动力
ejector *n.*	喷射器
elevation *n.*	立面
encapsulated *n.*	密封的
energy density	能量密度
envelope structure	围护结构
environmentally friendly	环保的
evapotranspiration *n.*	蒸散
expanded graphite powder	膨胀石墨粉
expanded natural graphite	膨胀天然石墨
extensive *adj.*	大量的
fatty acids	脂肪酸
feasibility *n.*	可行性
flat fins	平翅片
flat plate collector	平板集热器
forced circulation	强制循环
Fresnel lens	菲涅耳透镜
frost resistance	抗寒性
fusion *n.*	熔化
gas turbine engine	燃气轮机
glare *n.*	眩光
glass cover	玻璃罩
graphite flakes	石墨片
graphite *n.*	石墨
gravel–water heat storage	砾石-水蓄热
greenhouse effect	温室效应
greenhouse gas emission	温室气体排放
ground source heat pump	地源热泵
heat collection efficiency	集热效率
heat conduction oil	导热油
heat conducts	热传导
heat dissipation	散热
heat flux	热流密度
heat pump	热泵

heat sink	吸热部件
heat storage capacity	蓄热能力
heat storage tank	蓄热水箱
heat transfer medium	传热介质
heat transfer	传热
heliostats $n.$	定日镜
honeycomb $n.$	蜂窝型
hot water heat storage	热水蓄热
household systems	户式系统
humidification $n.$	加湿
humidifier $n.$	加湿器
hydrogen $n.$	氢气
hydrophobic distillation film	疏水蒸馏膜
in conjunction with	共同，配合
in sync	同时
inclination angle	倾斜角
indoor temperature	室内温度
infiltrating $n.$	渗透
infrared radiation	红外辐射
initial investment	初投资
inorganic $adj.$	无机的
insulating layer	绝缘层
insulation panel	隔热板
integral $n.$	整体
intermediate heat transfer medium	中间换热介质
irreplaceable	不可替代的
jet refrigeration	射流制冷
latent heat storage	潜热蓄热
latent heat	潜热
lattice type	花格式
light transmittance	透光率
linear Fresnel thermal power system	线性菲涅尔式热发电系统
liquid sodium	液态钠
long–term heat storage	长期蓄热
long–term storage	长时间储存
loosely $adv.$	宽松地
magnetron sputtering	磁控溅射
maintenance $v.$	维护

mechanical strength	机械强度
melting point	熔点
membrane *n.*	膜
microcapsule *n.*	微胶囊
mineral cotton *n.*	矿物棉
mineral oil	矿物油
miniaturized package	微型包装
mining *n.*	矿业
mirror field	镜场
molten salt	熔盐
multidisciplinary	多学科
nano – composite phase change materials	纳米复合相变储热材料
national standard	国家标准
natural circulation	自然循环
natural convection	自然对流
network structure	网状结构
non – combustible *adj.*	不易燃的
non – concentrating solar collector	非聚光型太阳能集热器
non – permeable *adj.*	不可渗透的
non – toxicity *n.*	无毒性
oil – fired boiler	燃油锅炉
operating temperature	工作温度
operation parameter	运行参数
optical system	光学系统
ordinary light	普通光线
organic *adj.*	有机的
oven *n.*	烤箱
overheating *n.*	过热
oxygen *n.*	氧气
parabolic trough solar thermal power system	槽式太阳能热发电系统
paraffin *n.*	石蜡
passive solar house	被动式太阳房
penetrate *v.*	穿透
permeable *adj.*	可渗透的
petrochemical *n.*	石化行业
phase change	相变
phase change heat storage	相变蓄热
phase change storage	相变储热

phase transition	相变
plastic film–insulated flat plate collector	塑料薄膜绝缘平板集热器
plate solar oven	平板太阳能灶
plumbu *n.*	铅
pollutants *n.*	污染物
polycarbonate *n.*	聚碳酸酯
polyester *n.*	聚酯纤维
polyethylene glycol	聚乙二醇
polyethylene *n.*	聚乙烯
polymer *n.*	聚合物
porous *adj.*	多孔的
prototype *n.*	模型
purity *n.*	纯度
radiation *n.*	辐射
receiving cavity	接收腔
reflectors *n.*	反射镜
refrigeration coefficient	制冷系数
refrigeration power	制冷功率
refrigeration *n.*	制冷
reliable *adj.*	可靠的
renewable energy	可再生能源
renewable *adj.*	可再生的
residual air	余气
residual *adj.*	剩余的
reversible chemical reaction	可逆化学反应
roof *n.*	屋顶
rotating parabolic mirror	旋转抛物面反射镜
rubber seal	橡胶密封
rural area	郊区，农村地区
rural areas	农村地区
scattered light	散射光
scattered radiation	散射辐射
schematic *n.*	示意图
sealed with	用…密封
seasonal *adj.*	季节的
secondary reflector	二次反射镜
selective absorption coating	选择性吸收涂层
selective absorption film	选择性吸收膜

sensible heat storage	显热蓄热
serpentine flow channel flat plate collector	蛇形流槽平板收集器
serpentine *v*.	蜿蜒、蛇行
shortcoming *n*.	缺点
short-term heat storage	短期蓄热
siphon *n*.	虹吸管
sloping *adj*.	倾斜的
Solar absorption refrigeration	太阳能吸收式制冷
solar agricultural drying	太阳能农产品烘干
solar direct flow hot water system	太阳能直流式热水系统
solar driven air-conditioning system	太阳能驱动空调系统
solar driven desiccant evaporative refrigeration	太阳能驱动蒸发式制冷
solar drying cooling refrigeration	太阳能干燥冷却制冷
solar energy high temperature thermal chemical reaction	太阳能高温热化学反应
solar forced circulation hot water system	太阳能强制循环热水系统
solar furnace	太阳炉
solar house	太阳房
solar natural circulation hot water system	太阳能自然循环热水系统
Solar seawater desalination	太阳能海水淡化
solar solid adsorption refrigeration	太阳能固体吸附式制冷
solar steam injection refrigeration	太阳能蒸汽喷射制冷
solar thermal energy	太阳热能
solar thermal power generation	太阳能热发电
solar thermal storage material	太阳能储热材料
solar thermal utilization	光热利用
solid sealing	固态封接
solid type	实体式
solidification *n*.	凝固
specific heat capacity	比热容
spring support	弹簧支架
stainless steel	不锈钢
stainless *adj*.	不锈钢的
Stirling engine	斯特林发动机
suction plate	吸热板
surrounding environment	周围环境
temperature fluctuation	温度波动
thermal efficiency	热效率

Appendix

thermal insulation curtains	保温窗帘
thermal resistance	热阻
thermal storage wall type	集热蓄热墙式
thermo – mechanical	热力性能
tiny pores	微孔
toughened glass	钢化玻璃
tower solar thermal power system	塔式太阳能热发电系统
traditional energy	传统能源
transmittance $n.$	透光率
transparent $adj.$	可穿透的
trial operation	试运行
trial $n.$	试验
trough solar collector	槽式聚光器
tubular collector	管状集热器
two – axis tracking	双轴追踪
ultrasonic $adj.$	超音速的
uneven distribution	不均匀分布
utilization rate	利用率
vacuum chamber	真空室
vacuum collector tubes	真空集热管
various $adj.$	各种各样的
veins $n.$	脉络、纹理
ventilation hole	通风孔
ventilation outlet	出风口
ventilation $n.$	通风
water wall type	水墙式
welded $v.$	焊接
wind resistance	风阻
window – wall ratio	窗墙比
working medium	工作介质

References

[1] HUACUZ J, URRUTIA M. Proceedings of the International Workshop Charge Controllers for Photovoltaic Rural Electrification Systems [M]. Cuernavaca: Electrical Research Institute, 1998.

[2] Robert Hull. Properties of Crystalline Silicon [M]. Stevenage: INSPEC, 1999.

[3] GREEN M, KEEVERS M. Optical properties of intrinsic silicon at 300K [J]. Progress in Photovoltaics, 1995, 3: 189-192.

[4] KOLODINSKI S, WERNER J, WITTCHEN T, QUEISSER H. Quantum efficiencies exceeding unity due to impact ionization in silicon solar cells [J]. Applied Physics Letters, 1993, 63: 2405-2407.

[5] CLUGSTION D, BASORE P. Modelling Free-Carrier Absorption in Solar Cells [J]. Progress in Photovoltaics, 1997, 5: 229-236.

[6] SPROUL A, GREEN M. Improved value for the silicon intrinsic carrier concentration from 275 to 375 K [J]. Journal of Applied Physics, 1991, 70: 846-854.

[7] PHILLIPS J, BIRKMIRE R, MCCANDLESS B, MEYERS P, SHAFARMAN W. Polycrystalline heterojunction solar cells: A device perspective [J]. Physica Status Solidi, 1996, 194 (b): 31-39.

[8] FAN J, PALM B. Optimal design of amorphous/crystalline tandem cells [J]. Solar Cells, 1984, 11: 247-261.

[9] CONTRERAS M, EGAAS B, RAMANATHAN K, HILTNER J, SWARTZLANDER A, HASOON F, NOUFI R. Progress toward 20% efficiency in Cu (In, Ga) Se_2 polycrystalline thin-film solar cells [J]. Progress in Photovoltaics, 1999, 7: 311-316.

[10] JASENEK A, RAU U, WEINERT K, KOTSCHAU I, WERNER J. Radiation resistance of Cu (In, Ga) Se_2 solar cells under 1-MeV electron irradiation [J]. Thin Solid Films, 2001, 387: 228-230.

[11] MITCHELL K, EBERSPACHER C, ERMER J, PAULS K, PIER D. $CuInSe_2$ cells and modules [J]. IEEE Transactions electronic devices, 1990, 37: 410-417.

[12] KAZMERSKI L, WAGNER S. Cu-Ternary Chalcopyrite Solar Cells [M]. London: Academic Press, 1985.

[13] HANEMAN D. Properties and applications of copper indium diselenide [J]. Critical Reviews In Solid State and Materials Sciences, 1988, 14: 377-413.

[14] ROCKETT A, BIRKMIRE R. $CuInSe_2$ for photovoltaic applications [J]. Journal of Applied Physics, 1991, 70: R81-R97.

[15] RAU U, SCHOCK H. Electronic properties of Cu (In, Ga) Se_2 heterojunction solar cells - recent achievements, current understanding, and future challenges [J]. Journal of Physics D-Applied Physics, 1999, A69: 131-147.

[16] SCHNITZER I. Ultrahigh spontaneous emission quantum efficiency, 99.7% internally and 72% externally, from AlGaAs/GaAs/AlGaAs double heterostructures [J]. Applied Physics Letters, 1993, 62 (2): 131.

[17] WANG X. Design of GaAs solar cells operating close to the Shockley-Queisser limit [J]. IEEE

References

Journal of Photovoltaics, 2013, 3 (2): 737.

[18] PULFREY L D. Photovoltaic power generation [M]. New York: Van Nostrand Reinhold Co., 1978.

[19] RIVERS P N. Leading edge research in solar energy [M]. New York: Nova Science Publishers, 2007.

[20] HOPPE H, SARICIFTCI N S. Organic solar cells: An overview [J]. Journal of materials research, 2004, 19 (7): 1924-1945.

[21] HALLS J J M, FRIEND R H, ARCHER M D, HILL R D. Clean electricity from photovoltaics [M]. London: Imperial College Press, 2001.

[22] CHIU S W, LIN L Y, LIN H W, CHEN Y H, HUANG Z Y, LIN Y T, LIN F, LIU Y H, WONG K T. A donor-acceptor-acceptor molecule for vacuum-processed organic solar cells with a power conversion efficiency of 6.4% [J/OL]. Chemical Communications, 2012, 48 [2011-12-23]. http://www.rsc.org/suppdata/cc/c2/c2cc16390j/c2cc16390j.pdf.

[23] LI B. Review of recent progress in solid-state dye-sensitized solar cells [J]. Solar Energy Materials and Solar Cells, 2006, 90 (5): 549-573.

[24] YELLA A, LEE H W, TSAO H N. Porphyrin-sensitized solar cells with cobalt (II/III) - based redox electrolyte exceed 12 percent efficiency [J]. Science, 2011, 334 (6056): 629-634.

[25] FREITAG M, TEUSCHER J, SAYGILI Y, ZHANG X. Dye-sensitized solar cells for efficient power generation under ambient lighting [J/OL]. Nature Photonics, 2017, 11: 372-378 [2017-05-01]. https://www.nature.com/articles/nphoton.2017.60.

[26] LEE M M, TEUSCHER J, MIYASAKA T, MURAKAMI T N, SNAITH H J. Efficient hybrid solar cells based on meso-superstructured organometal halide perovskites [J]. Science, 2012, 338: 643-647.

[27] KOJIMA A, TESHIMA K, SHIRAI Y, MIYASAKA T. Organometal halide perovskites as visible-light sensitizers for photovoltaic cells [J]. Journal of the American Chemical Society, 2009, 131: 6050-6051.

[28] NOH J H, IM S H, HEO J H, MANDAL T N, SEOK S. Chemical management for colorful, efficient, and stable inorganic-organic hybrid nanostructured solar cells [J]. Nano Letters, 2013, 13: 1764-1769.

[29] SHENG R, HO-BAILLIE A W Y, HUANG S J, CHENG Y B, GREEN M A. Four-terminal tandem solar cells using $CH_3NH_3PbBr_3$ by spectrum splitting [J]. Journal of Physical Chemistry Letters, 2015, 6: 3931-3934.

[30] HEO J H. Efficient inorganic-organic hybrid heterojunction solar cells containing perovskite compound and polymeric hole conductors [J]. Nature Photonics, 2013, 7: 486-491.

[31] BURSCHKA J. Sequential deposition as a route to high-performance perovskite-sensitized solar cells [J]. Nature, 2013, 499: 316-319.

[32] LIU M, JOHNSTON M B, SNAITH H J. Efficient planar heterojunction perovskite solar cells by vapor deposition [J]. Nature, 2013, 501: 395-398.

[33] MALINKIEWICZ O. Perovskite solar cells employing organic charge-transport layers [J]. Nature Photonics, 2014, 8: 128-132.

[34] BALL J M, LEE M M, HEY A, SNAITH H J. Low-temperature processed meso-superstructured to thin-film solar cells [J]. Energy & Environmental Science, 2013, 6: 1739-1743.

[35] ELHODEIBY A S, METWALLY H M B, FARAHAT M A, Performance anyalysis of 3.6 kw rooftop grid connected photovoltaic system in EGYP [J]. International Conference on Energy Systems and Technologies, 2011, 151-157.

[36] DARUL' A I, STEFAN M. Large scale integration of renewable electricity production into the grids

[J]. Journal of Electrical Engineering, 2007, 58 (1): 58-60.

[37] PEARCE J M. Expanding Photovoltaic Penetration with Residential Distributed Generation from Hybrid Solar Photovoltaic + Combined Heat and Power Systems [J]. Energy, 2009, 34: 1947-1954.

[38] MOSTOFI M, NOSRAT A H, PEARCE J M. Institutional-Scale Operational Symbiosis of Photovoltaic and Cogeneration Energy Systems [J]. International Journal of Environmental Science and Technology, 2011, 8 (1): 31-44.

[39] 黄裕荣, 侯元元, 高子涵. 国际太阳能光热发电产业发展现状及前景分析 [J]. 科技和产业, 2014, 14 (9): 54-56.

[40] 童家麟, 吕洪坤, 李汝萍, 关键. 国内光热发电现状及应用前景综述 [J]. 浙江电力, 2019, 38 (12): 25-30.

[41] 袁炜东. 国内外太阳能光热发电发展现状及前景 [J]. 电力与能源, 2015, 36 (4): 487-490.

[42] 张争, 夏勇. 太阳能光热发电的发展现状及前景分析 [J]. 长江工程职业技术学院学报, 2013, 30 (1): 24-26.

[43] 陈昕, 范海涛. 太阳能光热发电技术发展现状 [J]. 能源与环境, 2012 (1): 90-92.

[44] 邬明亮. 太阳能光热发电技术特性与经济性研究 [J]. 青海电力, 2019, 38 (2): 18-22.

[45] 肖强, 王红艳, 王金平. 太阳能光热发电现状及发展策略分析 [J]. 中外能源, 2016, 21 (10): 26-30.

[46] 李方方, 袁亚周, 吴怡. 太阳能光热发电现状及前景分析 [J]. 上海节能, 2016 (7): 397-399.

[47] 王光伟, 许书云, 韩蕾, 孙鸿波. 太阳能光热利用主要技术及应用评述 [J]. 材料导报, 2014, 28 (S1): 193-196.

[48] 许岩. 中国太阳能光热发电技术研究现状 [J]. 能源与节能, 2016 (6): 84-86.

[49] 张福君, 李凤梅. 综述太阳能光热发电技术发展 [J]. 锅炉制造, 2019 (4): 33-36, 46.

[50] KINCAID N, MUNGAS G, KRAMER N, et al. Sensitivity analysis on optical performance of a novel linear Fresnel concentrating solar power collector [J]. Solar Energy, 2019, 180: 383-390.

[51] 肖慧杰, 王凯, 刘欣颖. 浅析太阳能发电技术 [J]. 内蒙古石油化工, 2010, 36 (3): 93-94.

[52] 周纬. 太阳能热利用技术概况 [J]. 农业工程技术 (新能源产业), 2012 (10): 33-35.

[53] 殷志强. 探讨太阳能热利用发展 [J]. 太阳能, 2009 (6): 6-9.

[54] 赵金玲, 陈滨, 王永学, 陈翠英. 被动式太阳能加热系统动态热特性研究 [J]. 大连理工大学学报, 2008 (4): 580-586.

[55] 杨婧. 被动太阳能采暖地区适用技术类型分析 [D]. 西安: 西安建筑科技大学, 2020.

[56] 李金平, 王航, 王兆福, 黄娟娟, 王春龙. 甘南藏区太阳能主被动联合采暖系统性能 [J]. 农业工程学报, 2018, 34 (21): 1-7.

[57] 郑豪放, 季杰, 郭超, 赵东升, 魏蔚. 基于太阳能炕的主被动复合采暖建筑的实验研究 [J]. 太阳能学报, 2018, 39 (4): 940-945.

[58] 李金平, 王兆福, 王航, 黄娟娟, 王春龙. 严寒地区主被动太阳能协同采暖室内舒适度研究 [J]. 西安建筑科技大学学报 (自然科学版), 2019, 51 (4): 584-590.

[59] 阚德民, 高留花, 刘良旭. 主动式太阳能供暖技术发展现状与典型应用 [J]. 应用能源技术, 2016 (7): 43-49.

[60] 李婷, 康侍民, 陈静. 主动式太阳能供暖系统的研究现状综述 [J]. 制冷与空调 (四川), 2013, 27 (6): 611-615.

[61] 安玉娇. U型玻璃真空管太阳能集热器热性能研究与优化设计 [D]. 北京: 北京建筑工程学院, 2011.

[62] 黄俊鹏, 陈讲运, 徐尤锦. 平板太阳能集热器技术发展趋势 [J]. 建设科技, 2017 (4): 40-47.

References

[63] 王佩明, 王志峰, 费良斌. 全玻璃真空管太阳空气集热器 [J]. 可再生能源, 2004 (4): 28-29.
[64] 周小波, 蒋富林, 孙伟. 热管式真空太阳能集热管及其应用 [J]. 太阳能, 2011 (16): 52-58.
[65] 严军. 热管真空管集热器及太阳能热水系统 [J]. 可再生能源, 2008 (5): 68-71.
[66] 曹丽华, 张来, 姜铁熘. 蛇形流道太阳能平板集热器的数值分析 [J]. 东北电力大学学报, 2018, 38 (1): 43-48.
[67] 季杰. 太阳能光热低温利用发展与研究 [J]. 新能源进展, 2013, 1 (1): 7-31.
[68] 朱冬生, 徐婷, 蒋翔, 黄银盛, 漆小玲. 太阳能集热器研究进展 [J]. 电源技术, 2012, 36 (10): 1582-1584.
[69] 李彬. 我国太阳能集热器的现状及未来发展 [J]. 节能, 2017, 36 (3): 9-11.
[70] 于志. 多种太阳能新技术在示范建筑中的应用研究 [D]. 合肥：中国科学技术大学, 2014.
[71] FERNANDEZ A I, PAKSOY H, KOCAK B. Review on sensible thermal energy storage for industrial solar applications and sustainability aspects [J]. Solar Energy, 2020, 209: 135-169.